OCS Study
MMS 2003-030

Workshop on Deepwater Environmental Studies Strategy: A Five-Year Follow-Up and Planning for the Future

May 29-31, 2002

Editors

William W. Schroeder
Carolyn F. Wood

Prepared under MMS Contract
1435-01-02-CA-85035
by
The University of Alabama
Dauphin Island Sea Lab/MESC
Dauphin Island, Alabama

U.S. Department of the Interior
Minerals Management Service
Gulf of Mexico OCS Region

New Orleans
May 2003

DISCLAIMER

This report was prepared under contract between the Minerals Management Service (MMS), the University of Alabama, and the Dauphin Island Sea Lab. This report has been technically reviewed by the MMS, and it has been approved for publication. Approval does not signify that the contents necessarily reflect the views and policies of the MMS, nor does mention of trade names or commercial products constitute endorsement or recommendation for use. It is, however, exempt from review and compliance with the MMS editorial standards.

REPORT AVAILABILITY

Extra copies of this report may be obtained from the Public Information Office (Mail Stop 5034) at the following address:

U.S. Department of the Interior
Minerals Management Service
Gulf of Mexico OCS Region
Public Information Office (MS 5034)
1201 Elmwood Park Boulevard
New Orleans, Louisiana 70123-2394
Telephone Numbers: 1-504-736-2519
                    1-800-200-GULF

CITATION

Suggested citation:

Schroeder, W.W. and C.F. Wood, eds. 2003. Workshop on Deepwater Environmental Studies Strategy: A Five-Year Follow-Up and Planning for the Future, May 29-31, 2002. U.S. Dept. of the Interior, Minerals Management Service, Gulf of Mexico OCS Region, New Orleans, LA. OCS Study MMS 2003-030. 135 pp.

ABOUT THE COVER

The graphic depicts the dramatic increase in the production of both natural gas and oil in the deeper waters of the Gulf. In 1985, less than 10 percent of the oil and natural gas production in the Gulf was attributed to the deep waters of the Gulf. That number has dramatically increased to nearly 60 percent in 2001. Estimates are that the percentage will continue to increase into the future.

Workshop on Deepwater Environmental Studies Strategy:
A Five-Year Follow-up and Planning for the Future

May 29-31, 2002
New Orleans Airport Hilton
New Orleans, Louisiana

Workshops sponsored by the Minerals Management Service (MMS) promote the dissemination and exchange of information about topics of concern relative to planned or potential Outer Continental Shelf (OCS) activities. In April 1997 MMS sponsored a workshop on environmental issues surrounding deepwater oil and gas development in the Gulf of Mexico. This workshop brought together experts from other government agencies, academia, and industry that were tasked with identifying information needs in the physical, biological/ecological, and social/economic sciences as they related to industry's recent and rapid advancement into the deeper waters of the Gulf. These discussions were then used as guidance in designing the MMS Deepwater Strategic Studies Plan, composed of a series of deepwater studies.

With increased development of the deep waters of the Gulf of Mexico for oil and gas, MMS must consider possible impacts to both the natural and human socioeconomic environment. As industry gains more experience in deep waters of the Gulf and studies better explain the deepwater environment, MMS seeks to gather experts from academia, industry, and government sources for a Deepwater Workshop. A similar workshop held in April 1997 provided guidance to the MMS for deepwater issues, which led to focused environmental studies to fill much needed data gaps.

OBJECTIVE

The purpose of the workshop was to 1) review the current status of industry activity in deepwater; 2) discuss any new environmental and/or social issues that have come to light since the first workshop; and 3) to examine information syntheses and data gaps identified during objectives one and two.

ACKNOWLEDGMENTS

This meeting was a joint effort between the University of Alabama (UA), the Dauphin Island Sea Lab (DISL) and the U.S. Department of the Interior's Minerals Management Service (MMS). The success of this workshop was due to the individuals below. The professional dedication and skills contributed on behalf of this effort before, during, and after the meeting deserve a special note of thanks and appreciation.

Gary Goeke, Mary Boatman, James Ray – Program Co-Chairs

Bill Bryant, Mike Rex, Steve Murdock – Session Chairs

Robert Avent, Harry Luton, Sarah Tsoflias – MMS Session Co-Chairs

David Walker, Paul Siegele, Dan Allen – Industry Session Co-Chairs

Jeff Childs – Sub-Group Chair

Anne Jochens, Gil Rowe, Randy Edwards – Invited Technical Overview Speakers

Carolyn Wood, Randy Schlude, Jonathan Martin – UA/DISL Support Staff

Debbie Vigil – Exhibit Coordinator

Wayne Plaisance – Exhibit Production

Margaret Metcalf, T. J. Broussard, Connie Landry – MMS Workshop Support Staff

The Staff of the New Orleans Airport Hilton, Kenner, Louisiana

William W. Schroeder
Program Manager

TABLE OF CONTENTS

LIST OF FIGURES

LIST OF TABLES

LIST OF POWERPOINT PRESENTATIONS
PRESENTED AT THE WORKSHOP

(Click on the title to view the presentation.)

I. Background

Workshop on Deepwater Environmental Studies Strategy: A Five-Year Follow-up and Planning for the Future

**Robert Avent, Greg Boland, Alexis Lugo Fernandez,
Gary Goeke, Harry Luton and Sarah Tsoflias
U.S. Department of the Interior, Minerals Management Service**

Over the last decade significant deepwater oil and natural gas discoveries transformed the Gulf of Mexico (GOM) from a declining oil and gas province to one of world-class stature. The move in the early 80's away from targeting specific groups of outer continental shelf (OCS) blocks to be offered for lease to more open access, or what the U.S. calls area-wide leasing, set the stage for innovative companies to begin to test the new deepwater frontier. Having broad access also promotes economies of scale because an expanding infrastructure over time lowers the size of what constitutes an economic field by allowing smaller, otherwise marginal fields, to be produced in conjunction with existing development.

Rapidly improving deep drilling and exploration technologies, and discoveries of very large reservoirs has justified the higher costs of deep exploration and development. With this accelerating interest, leasing in waters deeper than 400 meters has proceeded at an unprecedented rate, with exploration and production following closely behind. Interest in deep waters has also been heightened by the passage of the Deepwater Royalty Relief Act in 1995 and the development of technologies that allow for cost-effective development beyond the continental shelf. In recent years, the oil and gas industry has leased tracts in depths greater than 3,000 m and has drilled wells in water depths greater than 2,000 m. The challenge to the Minerals Management Service (MMS) is to insure that the environmental information base is adequate to address management issues needed in the deep Gulf.

Industry's offshore trend has been paralleled by MMS-sponsored research of the deep sea in general and chemosynthetic communities and marine mammals in particular. Still by 1997, there clearly was a myriad of issues not yet addressed. Recognizing the magnitude of this task, MMS and Louisiana State University hosted a workshop on *Environmental Issues Surrounding Deepwater Oil and Gas Development* in the spring of 1997 (Carney 1998). Concurrently, the Deepwater Subcommittee of the OCS Advisory Board's Scientific Committee assisted MMS in identifying and prioritizing its deepwater information needs.

The following topics were identified as critical environmental issues:

- Socioeconomic effects of deepwater activity on ports and coastal support facilities;
- Storage, handling and discharge of chemicals and drilling muds;
- Location and avoidance of deepwater benthic communities;
- Potential of gas hydrates to cause problems in operations; and
- Characterization of fates and effects of deepwater blowouts or pipeline leaks.

A research strategy was developed to address these issues, with an ongoing commitment to continue to work with all stakeholders as the research program evolved. It was recognized early on that new issues would emerge and that collaborative efforts with industry and other governments would be necessary because MMS on its own could not entirely fund the necessary research and would benefit from the views and input of others. The MMS led an international effort with Norway and 23 oil and gas companies to learn more about the behavior of oil/gas spills at great water depths. A successful field experiment, known as "DeepSpill", was held in June 2000 off the coast of Norway, in which a deep spill was observed and characterized. This information will be most useful in calibrating and validating deep spill models presently under development that will be used in spill contingency planning and response. Concerning disposal of synthetic based muds, MMS is currently participating in a jointly funded study, with industry and the U.S. Department of Energy, to evaluate the effects on the seabed from the discharge of cuttings from drilling with synthetic based drilling fluids. The U.S. Environmental Protection Agency will use the information gathered in their general discharge permits.

Before the 1997 Deepwater Workshop

Prior to 1997, the MMS had funded studies to collect information about the ecology of the deep Gulf. Based primarily on accumulated trawl data and bottom photographs collected with the *R/V Alaminos* from the 1960's through about 1980, a synthesis report was prepared (Pequegnat 1983). However, the MMS still lacked systematic faunal data except for the megafauna, and this was only semiquantitative at best. The *Northern Gulf of Mexico Continental Slope Study* (Gallaway 1988) was a descriptive field study designed largely to test or confirm the depth zonation scheme suggested by Pequegnat. The ongoing continental shelf studies were conducted as a series of regional efforts based on regional physiological and zoogeographic character. But because the Gulf was believed to change monotonously with depth in the 1970's, it was believed that one large descriptive study with stations placed on transects and using consistent methods, would work for the deep-sea. (We did not yet appreciate the Gulf's geochemical and physiographic complexity.) Consisting ultimately of stations on a few shallow-to-deep and E-W transects, the study concentrated on benthic and benthopelagic sampling. Included were sediment characteristics; meio-, macro-, and megafauna; contaminants; and hydrography; with the usual comparisons of the non-biological variables and community structure.

During the latter phases of the *Northern Gulf of Mexico Continental Slope Study* (Gallaway 1988) field program (late 1984) that researchers recognized chemosynthetic species in dredge and trawl tows from an area known to have sediments containing oil, gas hydrates, and hydrogen sulfide. Carbon isotopic analyses confirmed that the tubeworms and molluscs lived via a chemosynthetic strategy (MacDonald et al. 1995). In 1995, the MMS initiated *Change and Stability in Gulf of Mexico Chemosynthetic Communities* to expand on the first study of chemosynthetic communities (MacDonald et al. 1995). The MMS recognized that there were still few data on (1) the ecological interactions in these communities, (2) temporal stability and change within the communities, and (3) the physicochemical habitats that support them. The second study addresses both nonbiotic habitat and biotic questions to understand the conditions necessary for community establishment and maintenance.

In other areas of oceanography, the learning curve was just as steep. While some information was available regarding the physical oceanography of the surface waters of the deep Gulf (SAIC 1989), nothing was known about the current regime near the bottom though, generally, it was believed to be quiescent. Only recently, through the use of 3-D seismic has the geology of the Gulf been more fully understood. Additionally, little was known about the socioeconomic impacts from offshore oil and gas activities along the Gulf coast.

The 1997 workshop consisted of four sessions: physical oceanography, ecology, sea floor geology, and socioeconomics. During the course of the workshop, two ad hoc sessions were formed; fisheries issues and deepwater operational process. As a result of the recommendations from the workshop, 30 studies have been funded (Table 1).

Table 1.
Completed and Ongoing Studies Funded in Response to the 1997 Deepwater Workshop

---

Completed Studies

Gulf of Mexico Deepwater Information Resources Data Search and Literature Synthesis
Deepwater Physical Oceanography Reanalysis and Synthesis of Historical Data
Offshore Data Search and Synthesis on Highly Migratory Species in the GOM and the Effects of Large Fish Attracting Devices (FADS)
Literature Review: Environmental Risks of Chemical Products Used in Deepwater Oil & Gas Operations
The Fate and Effects of Synthetic-Based Drilling Fluids and Associated Cuttings Discharged into the Marine Environment

Ongoing Ecological Studies

Summary of the Northern Gulf of Mexico Continental Slope Studies
Offshore Petroleum Platforms: Functional Significance for Larval Fish Across Longitudinal and Latitudinal Gradients
Assessment and Reduction of Taxonomic Error in Benthic Macrofauna Surveys: An Initial Program Focused on Shelf and Slope Polychaete Worms
Development of a Deepwater Environmental Data Model
Northern Gulf of Mexico Continental Slope Habitats and Benthic Ecology
Bluewater Fishing and Deepwater OCS Activity: Interactions Between the Fishing and Petroleum Industries in Deepwaters of the Gulf of Mexico
Effects of Oil and Gas Exploration and Development at Selected Continental Slope Sites in the Gulf
Sperm Whale Seismic Study (SWSS)

Table 1. Continued
Completed and Ongoing Studies Funded in Response to the 1997 Deepwater Workshop

Ongoing Physical Oceanography Studies

Observation of Deepwater Manifestation of Loop Current Rings
Deepwater Currents at 92 W
Study of Subsurface, High-Speed Current Jets in the Deep Water Region of the Gulf of Mexico
Analysis and Validation of a Mechanism that Generates Strong Mid-depth Currents and a Deep Cyclone Gyre in the Gulf of Mexico
Modeling and Data Analysis of Subsurface Currents on the Northern Gulf of Mexico Slope and Rise: Effects of Topographic Rossby Waves and Eddy-Slope Interaction
Study of Subsurface, High-Speed Current Jets in the Deep Water Region of the Gulf of Mexico
Exploratory Study of Deepwater Currents in the Gulf of Mexico

Ongoing Socioeconomic Studies

Labor Migration and the Deepwater Oil Industry in Houma
An Analysis of the Socioeconomic Effects of OCS Activities on Ports and Surrounding Areas in the Gulf of Mexico Region
Labor Migration and the Deepwater Oil Industry
Potential Spatial and Temporal Vulnerability of Pelagic Fish Assemblages in the Gulf of Mexico to Surface Oil Spills Associated with Deepwater Petroleum Development
Assessing and Monitoring Industry Labor Needs
Benefits and Burdens of OCS Deepwater Activities on Selected Communities and Local Public Institutions
OCS-Related Infrastructure in the Gulf of Mexico
Supply Logistics of OCS Oil and Gas Development in the Gulf of Mexico -- Evaluation of Technological and Economic Parameters of Ports as Supply and Manufacturing Bases

Other Ongoing Studies

Understanding the Processes that Maintain the Oxygen Levels in the Deep Gulf of Mexico
Joint Industry Project, "Gulf of Mexico Comprehensive Synthetic Based Muds Monitoring Program"

What is Known Today

Research over the past five years has greatly increased our understanding of the environment of the Gulf of Mexico. What was once thought to be a simple, monotonous system is now known to be complex and dynamic both above and below the seafloor. The following sections discuss some of what has been learned since the last workshop. More specific information was presented in the plenary session and the details can be found in the final reports.

Geology

Understanding the Gulf of Mexico environment begins with the complex geologic setting, which creates a foundation for the life within and above. The majority of the complex geology of the Gulf of Mexico is due to horizontal and vertical movement of salt. The Louann Salt, which underlies the sedimentary provinces, was deposited in Jurassic time in the basin during repeated episodes of flooding and evaporation of shallow saline waters. Bathymetric maps of the continental slope in the northwestern Gulf of Mexico (Bryant et al. 1990; Bouma and Bryant 1995) reveal the presence of over 105 intraslope basins with relief in excess of 150 meters. The southern edge of the salt mass within the northern Gulf is the Sigsbee Escarpment. The lower continental slope contains numerous submarine canyons that cut into the Escarpment. Submarine fans of various sizes extend seaward of the canyons onto the continental rise. On the upper continental slope, gas hydrates and underconsolidated gassy sediments are common.

The geology of the deepwater Gulf of Mexico influences the geochemical and biological processes in the region. Faults act as migration conduits for subsurface fluid and gas to the seafloor. Interesting benthic habitats are created due to this fluid and gas seepage and chemosynthetic communities are common. Understanding the geology of the deepwater continental slope aids in understanding the deepwater biologic communities.

Due to the interaction of halokinesis and rapid sedimentation, several geologic hazards, or geohazards, exist in this region. Salt movement can cause potential hazards such as seafloor fault scarps, slumping from steep unstable slopes, shallow gas pockets, seeps and vents, and rocky or hard bottom areas. Gas hydrate instability could result in the dissociation of hydrates and subsequent sediment slumps and slides. High sedimentation rates cause unconsolidated, high-water-content, and low-strength sediments. Under these conditions, the sediments can be unstable, and slope failure or mass transport of sediments can result. Rapid sediment deposition and salt movements create instabilities by over steepening slopes resulting in slumps, creep and debris flows.

Physical Oceanography

Above the seafloor, the circulation of water within the Gulf helps create communities through the transport of nutrients, plankton, and larvae. Currents are also important because they transport the discharges from oil and gas activities. As an initial step in understanding the deepwater physical oceanography, MMS funded a synthesis of the present "state of knowledge" (Nowlin et al. 2001). The data compiled in this study represents the most complete and updated source of information on deepwater physical oceanography of the Gulf. Examination of the spatial and temporal distribution of the data reveals a number of data gaps. Except in the central Gulf, there are few current data sets in water depths of 1000 m or more, and few stations available south of 25°N.

The circulation within the Gulf of Mexico is driven principally by two sources of energy. The main source consists of the Yucatan Current and other circulation features that enter the Gulf from the Caribbean Sea through the Yucatan Channel and forms the Loop Current. The second major energy source is wind stress forcing. Other effects include episodic currents forced by high frequency but strong atmospheric events including tropical cyclones, extratropical cyclones, cold air outbreaks, and other frontal passages. Thermohaline forcing is known to be important over the Gulf shelves, e.g., buoyancy forcing by river discharge affects the nearshore coastal currents over the shelves. However, no thermohaline forcing of consequence or significant water mass formation are known to occur in the deepwater Gulf.

Ecology

Arguably, the most recognized studies of the MMS Environmental Studies Program have been the "Marine Ecosystems" studies. Almost as soon as the National Environmental Policy Act (NEPA) was passed and there was an Environmental Studies Program (1973), the Bureau of Land Management (the BLM, a predecessor of the MMS) began to fund a series of ambitious descriptive regional field studies around the Gulf. In some cases, these continental shelf studies were the first truly systematic, multidisciplinary, multiyear studies of a given area. Since then, there have dozens of final reports on shelf ecosystems. In the late 1970's it was already apparent to the BLM that the industry was moving onto the upper continental slope, and early planning was in order for studies of the continental slope.

Deep-sea animals live mostly under conditions of total darkness, low temperature, nearly featureless mud, and sparse food resources. They are generally small and fragile and the structure of benthic communities is generally well known, if not satisfactorily explained, ecologically. Many of the same sampling and statistical methods used in shallow water apply well in the deep-sea environment. However the costs for sea time and sample treatments are usually significantly higher. The character and rarity of many deep animals, the high species diversity, and the relative paucity of qualified systematists add to the difficulties.

In 1988 when the *Northern Gulf of Mexico Continental Slope Study* final report was received (Gallaway 1988), the MMS was already aware that the classical picture of the continental slope was changing. Many newer scientific findings were made, among the most important being the discoveries of productive chemosynthetic communities in 1984, and the collection of far better information on geological structural complexity and related biogeochemical processes. It was now generally conceded that the available information on the continental slope was inadequate.
During the 1997 workshop, the discussion group on deep-sea ecology focused on the state of knowledge of benthic communities, its composition and complexity, associated biogeochemical processes, habitat heterogeneity and trophic considerations. Several studies were funded based on the recommendations including *Northern Gulf of Mexico Continental Slope Habitats and Benthic Ecology* (in progress), *Management Applicability of Contemporary Deep-Sea Ecology and Reevaluation of Gulf of Mexico Studies* (Carney 2001) and *Effects of Oil and Gas Exploration and Development at Selected Continental Slope Sites in the Gulf of Mexico* (in progress).

Fisheries

Deepwater fisheries as they relate to offshore oil and gas development have only recently become an issue as a result of the shift of interest from the continental shelf to the deeper waters of the northern Gulf of Mexico. The potential for interactions between bluewater fishing and the deepwater energy industry was raised as a concern by a fisheries subcommittee during the first deepwater workshop sponsored by MMS (Carney 1998). This concern provided the impetus for the recently completed study, *Bluewater Fishing and OCS Activity, Interactions Between the Fishing and Petroleum Industries in Deepwaters of the Gulf of Mexico* by Continental Shelf Associates (CSA 2002, submitted). This study addressed conflicts between pelagic longliners and geophysical survey vessels, longline interactions with other surface operations, and bottom trawl interactions with structures. A second issue raised during the workshop was the potential for deepwater petroleum structures (DPSs) to impact Gulf fish and fisheries by acting as fish aggregating devices (FADs). A recently released report prepared by the United States Geologic Survey *Potential for Gulf of Mexico Deepwater Petroleum Structures to Function as Fish Aggregating Devices (FADs) – Scientific Information Summary and Bibliography* (Edward and Sulak 2002) discusses the topic.

Socioeconomics

The socioeconomic breakout group for the 1997 deepwater workshop (Carney 1998) faced a problem similar to that faced by earlier reviews and workshops; the lack of studies to review. Until the 1990s, the GOMR had conducted few social and economic studies; those that were funded addressed a narrower range of issues. By 1990, external reviews, advice from the agency's Science Committee, and the traumatic 1980s oil price bust had convinced the MMS of the need to expand its consideration of the program's social and economic effects and several new socioeconomic studies were underway. All of these addressed the consequences of the oil-price collapse; most of them had employed a "boom/bust" model to explain its social consequences. However, while these studies show the growing emphasis on socioeconomics, the MMS's present-day approach to social and economic research for the Gulf Region developed later, beginning in 1992. That year the NRC reviewed the MMS studies program (NRC 1992), the GOMR held its socioeconomic agenda setting workshop (Gramling and Laska 1993), and the GOMR studies program began to incorporate their findings into its studies planning process.

Nevertheless, five years later when the 1997 GOMR deepwater workshop was held, its members still confronted an almost empty slate. While the bust-related studies were complete, only two published studies reflected the initial MMS response to the NRC review and GOMR workshop. One was an expanded version of a workshop "white paper" that linked changes in offshore petroleum technology and business practices to social effects; the other was an assessment of stakeholder issues for the central Gulf. Several others were in draft or underway. The agency's slow response to the NRC and workshop recommendations was due in part to the time needed to reshape the studies planning process. However, the Gulf also faced a budget problem. The NRC reviews of physical oceanography and ecology had been first out of the gate. By 1997, much planning had been completed but most studies funds had already been committed to addressing NRC suggestions in these areas.

Now, however, the 2002 deepwater workshop does not face an empty slate. The last decade of MMS Gulf socioeconomic studies has been robust and varied, particularly when compared to earlier times. Many of these studies are now published, many others will soon conclude. Guidance from the 1992 NRC review (NRC 1992) and workshop (Gramling and Laska 1993) and the 1997 deepwater workshop (Carney 1998) have substantially shaped this output—at least as this guidance has been interpreted and operationalized by MMS.

The timing of the GOMR 2002 deepwater workshop is fortuitous because this is the time to retool for the second wave of research. Part of this task is to assess what the first wave left undone, what needs following up, what failed, and what did not. To facilitate this process, Drs. Steve Murdock, Stan Albrecht, and F. Larry Leistritz have prepared an assessment of MMS Gulf socioeconomic study reports completed during the last decade and MMS has prepared a description of the ongoing studies. The MMS is also providing a history of the MMS Gulf socioeconomic studies program that was prepared for its Scientific Committee.

This workshop is intended to consider what MMS should do in light of what its research has accomplished. The participants may identify specific information needs and specific studies that would address them. However, MMS is not seeking a list of suggested study titles but, rather, what might best be characterized as "strategic thinking" about issue selection, methodologies, and research and socioeconomic assessment goals. It seeks discussion on such questions as: Where should the next round of study effort be concentrated and why? What areas, questions, and problems are most critical for an environmental assessment? What might be the most effective ways to research them? What obviously important questions might not be effectively addressed? To facilitate this discussion, MMS will provide a hypothetical research agenda that identifies problem areas in the current research program and the goals and methodologies of a series of studies that might address them.

What's Next?

Five years have passed since the 1997 workshop (Carney 1998) and it is time to examine where the program has been and where it is going. This workshop is structured to parallel the first deepwater workshop with the exception of physical oceanography. A separate workshop was held to discuss deepwater physical oceanography in September, 2000 (McKay et al. 2001). Therefore, this workshop was broken into three main sessions; ecology, sea floor geology, and socioeconomics. One ad hoc group was formed to discuss fisheries.

The remoteness of deepwater development has brought about consideration of several non-routine issues. The MMS has had discussions with various companies leading efforts to investigate offshore support facilities for equipment and consumables, temporary housing of personnel, emergency landing facilities for aircraft, field hospitals, offloading terminals, central gathering facilities, shuttle tanker transport of produced hydrocarbons, waste management, and mariculture initiatives. With each new proposal, environmental and technological challenges emerge. The role MMS will play in these nonroutine initiatives designed to support deepwater development will evolve as individual projects become more certain. Also, given the unknown aspects of what awaits discovery in ever-deeper waters, we must be environmentally vigilant. The recent discoveries of chemosynthetic communities, deepwater furrows on the seabed, and "iceworms" living on methane hydrates may be just a preview of what is to come. It is incumbent on us to use emerging technology being developed, such as remotely operated vehicles and ever-improving sensors, to further our understanding of these deep resources. A challenge for our future is to be ready and able to assess, protect, and possibly use these resources in a sustainable manner.

References

Bouma, A. H. and W. R. Bryant. 1995. Physiographic features on the northern Gulf of Mexico continental slope. Geo-Marine Letters 14:252-263.

Bryant, W. R., J. R. Bryant, M. H. Feeley and G. S. Simmons. 1990. Physiography and bathymetric characteristics of the continental slope, Gulf of Mexico. Geo-Marine Letters 10:182-199.

Carney, R. S. 1998. Workshop on environmental issues surrounding deepwater oil and gas development. Final Report. U.S. Department of the Interior, Minerals Management Service, New Orleans, LA. OCS Study MMS 98-0022.

Carney, R. S. 2001. Management Applicability of Contemporary Deep-Sea Ecology and Reevaluation of Gulf of Mexico Studies. U.S. Department of the Interior, Minerals Management Service, Gulf of Mexico OCS Region, New Orleans, LA. OCS Study MMS 2001-095. 174 pp.

Continental Shelf Associates, Inc. 2000. Deepwater Gulf of Mexico Environmental and Socioeconomic Data Search and Literature Synthesis. Volume I: Narrative Report. OCS Study MMS 2000-049. U.S. Department of the Interior, Minerals Management Service, Gulf of Mexico OCS Region, New Orleans, LA. 340 pp.

Continental Shelf Associates. 2002 submitted. Deepwater Program: Bluewater fishing and OCS activity: Interactions between the fishing and petroleum industries in deepwaters of the Gulf of Mexico. Draft Final report to). U.S. Department of the Interior, Minerals Management Service, contract #1435-01-99-CT-31011. 185 pp. + appen.

Edwards R. E and K. J. Sulak. 2002. Potential for Gulf of Mexico deepwater petroleum structures to function as fish aggregating devices (FADs) – Scientific information summary and bibliography. Final Project Report, U.S. Department of Interior, Geological Survey, USGS BSR 2002-0005 and Minerals Management Service, Gulf of Mexico OCS Region, New Orleans, Ala, OCS Study MMS 2002-39. 261 pp.

Gallaway, B. J. (Ed.). 1988. Northern Gulf of Mexico Continental Slope Study, Final Report: Year 4. Volume II: Synthesis Report. Final report submitted to the Minerals Management Service, New Orleans, LA. OCS Study/MMS 88-0053. 318 pp.

Gramling, R. B. and S. B. Laska. 1993. A social science research agenda for the Minerals Management Service in the Gulf of Mexico. Prepared by Louisiana Universities Marine Consortium, University Research Initiative for the U.S. Dept. of the Interior, Minerals Management Service, Gulf of Mexico OCS Region, New Orleans, LA. OCS study MMS 93-0017. May. ix, 62 pp.

MacDonald, I. R., W. W. Schroeder and J. M. Brooks. 1995. Chemosynthetic Ecosystems Studies, Final Report. U.S. Dept. of the Interior, Minerals Management Service, Gulf of Mexico OCS Region, New Orleans, LA. OCS Study MMS 95-0023. 338 pp.

McKay, M., J. Nides, L. Atkinson, A. Lugo-Fernandez and D. Vigil. 2001. Workshop on the physical oceanography slope and rise of the Gulf of Mexico, September, 2000. U.S. Dept. of the Interior, Minerals Management Service, Gulf of Mexico OCS Region, New Orleans, LA. OCS Study MMS 2001-021. 151 pp.

National Research Council (NRC). 1992. Assessment of the U.S. outer continental shelf Environmental Studies Program: III. Social and economic studies. Washington, DC: National Academy Press, National Academy of Sciences. x,164 pp.

Nowlin, W. D., A. E. Jochens, S. F. DiMarco, R. O. Reid and M. K. Howard. 2001. Deepwater Physical Oceanography Reanalysis and Synthesis of Historical Data: Synthesis Report. U.S. Dept. of the Interior, Minerals Management Service, Gulf of Mexico OCS Region, New Orleans, LA, OCS Study MMS 2001-064. 528 pp.

Pequegnat, W. E. 1983. The Ecological Communities of the Continental Slope and Adjacent Regimes of the Northern Gulf of Mexico. U.S. Dept. of the Interior, Minerals Management Service, Gulf of Mexico OCS Region, New Orleans, LA, Contract No. AA851-CT1-12, 398 pp.

Science Applications International Corporation (SAIC). 1989. Gulf of Mexico Physical Oceanography Program, Final Report: Year 5. Volume II: Technical Report. U.S. Dept. of the Interior, Minerals Management Service, Gulf of Mexico OCS Region, New Orleans, LA, OCS Study MMS 89-0068. 333 pp.

II. Industry Perspectives

Going Beyond

David Walker
BP America Inc., Houston Texas

BP has a huge interest in the future of the deepwater Gulf of Mexico. Of the major operators, we find ourselves the clear leader in terms of resources with over 3 billion barrels of oil equivalent discovered to date. This makes the deepwater Gulf of Mexico the biggest single investment opportunity we have in our upstream portfolio. The safe and environmentally sound development of these resources is absolutely central to the future of BP. Your invitation to speak today asked for an operator's perspective on future environmental and social issues in the deepwater Gulf. Before I discuss that, I'd just like to say a little bit about BP, about how it feels to me as an engineer working there, and how environmental concerns impact what we do.

Three months ago I was invited to attend a two day workshop in London, addressing the subject of the year 2015 – specifically, what sort of company would BP be by then. It was, I think, the most interesting two days I have spent with the company in the past 30 years. Sixty people were present, 30 senior staff from BP and 30 external contributors, mainly academics from the US and UK. Environmental and social issues dominated the agenda. One of the most interesting talks at the workshop was given by an architect. He was a man whose views on society's relationship to the environment could be described as radical, at least by the standards of contemporary American society. Before he stood up to address us, I had already decided he wasn't from BP – there was something about his black suit, black shirt and large floppy black bow tie that marked him as 'not one of us'. But his message was hugely powerful and totally universal. He talked about toxicity and questioned why we tolerate so much in our lives. He explained how a new approach to building design could transform the impact they have on their natural surroundings – and he didn't just mean the aesthetic impact, but all aspects, including, in particular, their energy consumption. I found out later that his architectural partnership has been engaged by BP to consult on questions of design. Not just product design, but on the design of the buildings and platforms we construct, looked at from a radical environmental perspective. I mention this, not because I believe we are going to change the way we do our business overnight, but because it is symptomatic of the debate going on inside BP about where we want to get to as a company, in terms of our relationship with society and the natural environment.

This thinking starts at the top with our leadership. John Browne, our CEO, has summed it up very simply with three bullet points. You must admit it is commendably simple, direct and unambiguous. We're not there yet, but it clearly sets the direction for all our operations and if these things are important to the company leadership, then they get conveyed to staff and thus become a focus and there's plenty of evidence that it's working. Over the last 10 years our safety record has been transformed and now we are in the top quartile of our peers.

This is where I get the title of my talk – the sense that as a company we are not simply driven by regulatory compliance – but by a desire to go beyond that – to be true to our aspiration of no damage to the environment. For a major oil company this generates a tough paradox – how is it possible to be both a provider of cheap and efficient energy and a force for improving the environment? Let me briefly summarize where I see the debate going on inside BP.

We are paying close attention to what various international conventions are saying about these increasingly important issues for the planet. Learning how to determine the right answer when some of these priorities come into conflict is not easy. We may for example come across a technology with great scope for improving water quality, but if it is hugely inefficient in terms of energy consumption, we may have second thoughts about deploying it. The whole question of how we address our holistic impact on the environment and learn to make the right trade-offs is the big debate inside BP today. Our style is to listen to what the world is saying and to learn by doing.

I think we would all agree that, in general, the industry has a great track record in terms of devising financial metrics. Greenhouse gas emissions data as environmental performance indicators are evolving quickly. Social performance metrics are also beginning to mature. However if we think about an issue like biodiversity, one can argue that performance measures are still in their infancy. This presents a real challenge to us if we are going to incorporate actions to address biodiversity into the performance driven culture of our company. Okay I hear you say, all this is fine, but how do these high ideals translate into the real world of deepwater oil and gas production here in the Gulf of Mexico? Let me give you some examples. A cornerstone of our environmental policy is the need for measurement, transparency and continuous improvement. The way we have chosen to do this is via adoption of ISO 14001 certification.

We have been running this program here in the Gulf of Mexico since December 1998 and we were the first operator to introduce it. Twice a year we call in an independent authority to audit our business units. This entails a detailed review of all our measured environmental impacts – emissions to air, discharges to water, waste disposal, hydrocarbon spills and environmental fines. This data is published annually – evidence of what I mean by transparency. The audit also includes interviews with key personnel to gather evidence that our performance targets are understood and that a rigorous system is in place for their delivery. In the same way that we monitor safety performance, not just within our own operations but also within those of our contractors, so too with ISO 14001 compliance. Since 1998, we have certified all of our drilling rigs under long term charter. ISO 14001 embodies the concept of continuous improvement – each year the target levels on environmental impact measures are reduced. What has been achieved is truly remarkable.

For example, in 1998, BP announced that it would reduce greenhouse gas emissions by 10% from a 1990 baseline figure by 2010. This was to be accomplished against an aggressive production growth target for the company over this period, which effectively meant a 33% reduction in emissions per unit of production. This target was achieved this year – eight years ahead of schedule - and a new target has been announced to take us through to 2012. One of the important tools used by the company to meet the target was the establishment of an emission trading system between business units. This acted as a mechanism for identifying where in the company emission reductions could be achieved most economically. In comparison to other parts of the company our deepwater production facilities are amongst the lowest emitters of greenhouse gas – roughly 10% of the company average, expressed in terms of tons of CO_2 emitted per 1000 barrels produced. This whole approach to greenhouse gas emissions is not just good environmental practice – it also makes sound business sense. A recent analysis of the value of the steps we had taken to reduce emissions over the past few years demonstrated a net positive NPV to BP of several hundred million dollars.

All sorts of different technologies can contribute to achieving this. Last year we were the first company to moor a drilling rig using polyester moorings in a water depth where conventionally in the past we would have used dynamic positioning. This reduced fuel consumption and hence CO_2 emissions by over 60%. Similar levels of improvement are being achieved with drilling fluid discharges. For many years we've been working in collaboration with the MMS, the EPA, and our industrial partners to define new standards on what is permissible. As a result of the new EPA western GoM discharge permit, the implementation of a program tackling discharges of synthetic based drilling fluids will achieve a 50% reduction. It is mandatory within the company that every new development project generates an environmental strategy that details how the company's goals are being met. As mentioned earlier, this involves making difficult decisions to balance our goal of "no damage to the environment" with other important considerations such as safety, technical feasibility, reliability and, of course, cost. The strategy is subject to intense peer review and is reviewed at a senior level in the company. I hope I've managed to convey a sense of how a structured and audited process for meeting environmental targets is having a profound impact on our operations. It constitutes nothing less than a complete cultural change compared to the past. I touched on the subject of biodiversity earlier in the talk and I'd like to return to it.

I don't know how industry observers decide what is important to oil companies, but one simple way is to read their annual reports. We are certainly not the only major to address it there. I have a feeling that if you were to go back just a very few years, that would not have been the case. In September of 2000, BP, in collaboration with the University of Southampton Oceanography Centre, published a briefing paper on Deep-Sea Biodiversity and Hydrocarbon Exploration and Production. The paper acknowledges that not only is marine biodiversity becoming increasingly important, but that our level of understanding is very mixed. In some deepwater basins (and I would include the GoM in that group) we have a reasonable amount of data – while in other deepwater areas of interest to us we have very little. This is a huge challenge for us all and presents a tremendously fertile arena for collaboration between scientists and oil companies. Not only do our offshore facilities represent ideal sites for studies, but much of the technology we are developing to support our deepwater operations – such as ROV's and AUV's – is ideally suited to support such fields of scientific enquiry. We are already very active in this field. We have recently set up a collaborative program of scientific research in the fields of marine taxonomy and zoogeography at Texas A&M and Southampton Universities. The broad objectives are: provide support on the taxonomy of key benthic invertebrate groups; realize added scientific value from available BP material including samples; and provide a vehicle for broader international scientific collaboration.

Our contributions in this area in 2001 totaled over $1mm. I don't want to list all the donations, but let me just give you a couple of examples. We are a major supporter of the Gulf of Mexico Foundation which co-ordinates research on offshore platforms in conjunction with local universities. This included a conservation program for migratory waterfowl. We've dedicated 75,000 acres of wetlands and migratory waterfowl habitat to form a new wildlife refuge at Whitelake in Louisiana. We've also made a grant to Nicholls State University to test the designs of artificial reefs made with recycled materials. In recognition of this work, I am proud to say that BP was awarded the Gulf Guardian Award in 2002.

A final comment on biodiversity. It's not just for us a subject where we fund collaborative academic research. It is having a real impact on our operations. By the end of 2003, all of our key facilities around the world have to have biodiversity action plans in place. So once again we witness the company trying to "go beyond" - that is seeking to take a distinctive lead on an environmental issue in a practical way. One of the topics I was asked to address today concerned "Challenges for the Future". I guess I could set about presenting a list of technical topics and we can all think of a few potentially contentious areas. But rather than do that, I thought I would just talk about a piece of work we have carried out here in the Gulf, which for me at least, captures many of the positive features that we would like to see more of in the future.

Go back to 1997, and we found ourselves needing to know about the Sigsbee Escarpment, a 2,500' foot high, submerged steep slope, that marks the southern edge of the salt mass. We wanted to drill adjacent to the slope, and this raised a number of questions. We needed to understand the effect of drilling on the slope – whether for instance it might trigger any movements. We needed to understand the currents in the region to be able to safely moor and drill from our rigs. Should we make a discovery, we needed to be able to address long term production questions – for example – how would we route an export pipeline across this rugged subsea terrain? Our first problem concerned the limitations of conventional deep tow marine survey equipment in working over a steeply sloping, rugged deepwater escarpment. It is difficult to maintain the towed data recorder at a predetermined height off an undulating seabed. In deepwater, it is also difficult to follow a predetermined survey route, since the recorder can be more than a couple of miles distant from the towing vessel. It became clear that we needed to take a completely new approach and so in 1999, we brought over from Norway an Autonomous Underwater Vehicle or AUV.

Following successful sea trials, we became the first operator to use this technology here in the Gulf, and it is not an exaggeration to claim that it is has revolutionized deepwater surveying. In fact I would argue that without this step change in surveying technology, we could never have gathered such high quality data about the Sigsbee, and certainly not so quickly.

The shear volume of data coming back from the free swimming AUV presents new challenges – and opportunities – in terms of data processing and management. For the first time we can take advantage of 3D visualization techniques, to give ourselves an unprecedented perspective of seafloor morphology and processes and it wasn't just interesting data about seabed morphology which was collected.

I'm sure you all read about how the AUV was able to locate the wreck of a WW II submarine that was found in the Gulf last year. From the overlay of the images it is quite clear what we are looking at. But this breakthrough in seabed surveying was only one piece in a four-year program to improve our understanding of the escarpment.

What soon became clear was that it required a multi-disciplined approach – a complete integration of several applied sciences – geology and geophysics (we wanted to understand how movements of the salt masses in the seabed might be affecting slope stability), oceanography, geotechnics and marine biology – all were of huge importance in allowing us to address the

14

impact that our operations might have on the slope and vice versa. So the challenge here is one of integration – achieving a true marriage of the earth and life sciences. What we have taken away from this 4-year exercise are some salutary lessons on complexity – it isn't easy to understand what's going on in 6000' of water. We needed to bring in technology and learning from other deepwater basins where we're active, e.g., offshore Norway and the Caspian. But we believe that in completing this program of study on a part of the Sigsbee, we have set new standards for the industry in terms of due process and data gathering. Clearly this will be important as we move to new areas of the Gulf.

What I haven't covered in this talk is what sort of new production technology we might expect to see deployed in the Gulf in the future. That was a deliberate omission, because fundamentally I don't think it will be hugely different from what we see today. In the background of this slide you can see an artist's impression of our newest development, Thunderhorse. What you see is a large floating semi submersible, permanently moored over clusters of subsea wells. My guess is that we'll see a move towards more subsea production, leading eventually to seabed processing. But we'll still see a variety of hull types in use. I doubt if the basic concepts will change very much.

So to sum up, I hope I've given you at least a feel for how a major operator like BP approaches environmental issues here in the Gulf. I've talked about how our high level aspirations get translated into actions at Business Unit level. I've talked about our rapidly evolving agenda in biodiversity which is creating an exciting new field for collaborative work between industry, academia and regulatory bodies and I've given you some ideas of how we took on the challenge of determining how we could safely operate in a new area of the Gulf. All of this only works with a high degree of trust and openness between the interested parties. We have a great record of working in partnership with the MMS and academia. Long may it continue.

Going Beyond
Click on the title to view the presentation.

Workshop on Deepwater Environmental Studies Strategy: Industry Perspective

Paul Siegele
ChevronTexaco, New Orleans, Louisiana

The deep water Gulf of Mexico is one of the premier exploration plays available to the petroleum industry. Over 400 wildcat wells have been drilled in deep water since 1975. About 130 discoveries have resulted from the drilling for an overall success rate of slightly less than 1:3. The discoveries have an estimated ultimate recovery of over 11 billion barrels of oil equivalent.

The first offshore licensing in the Gulf of Mexico was offered by the State of Louisiana in 1936. Federal licensing in the shelf waters began in 1954. The move to deep water exploration started in 1983, and by 1991 approximately 2400 blocks were under lease. Exploratory drilling resulting from the early leasing activity led to the major discoveries at Mars, Ursa, Ram-Powell, Auger, and Brutus. The significant move to explore the ultra deep water began in 1995, and between the years 1995 and 1998 approximately 3200 additional blocks were awarded to industry at a bonus cost of $2.5 billion.

Exploratory drilling over the late 1990s has steadily climbed, from approximately 20 rank wildcat wells drilled in 1995 to over 60 rank wildcat wells drilled in 2000.

The technical challenges to be overcome, in addition to water depth, are due to the fact that almost the entire deep water area is covered by thick tabular salt. This salt layer creates seismic imaging problems as source energy has difficulty penetrating the salt and reflecting images located beneath the salt. The salt also creates additional challenges when converting to the recorded time into depth relations that can be mapped. The industry response to these imaging problems has been widespread acquisition of 3D datasets. In 1995 only about 30% of the deep water area was covered by 3D surveys. Today virtually the entire deepwater north of the abyssal plain is covered by speculative 3D surveys.

Advanced processing is required to extract useful depth information from the seismic. Employing a technique referred to as pre-stack depth migration, geophysicists are able to account for the changes in velocity in the sediments above and below salt with the velocities in the salt. The process is iterative, time consuming and expensive, but it is the only way to obtain clear pictures of geologic structures beneath the salt.

The salt also creates drilling challenges. Exploratory wells in the ultra-deepwater typically reach depths below 25,000'. It is difficult to predict pore pressure in the sediments beneath salt. Well costs can be typically between $30-40 million, and can exceed $100 million in extreme cases when things go wrong. The water depth and target drilling depth are the cause of drilling costs. Special drilling vessels capable of reaching such targets can cost $300,000 to $400,000 per day to operate.

Despite the challenges, the rewards can be great. The exploration drilling resulting from the latest licensing peaks of the late 1990s have led to several major discoveries that will soon be on production. Thunderhorse, Atlantis, Mad Dog and Tahiti have all been announced as discoveries of several hundred million barrels to over a billion barrels in size.

More typical discoveries are of a much more modest size. Of the 140 deep water discoveries announced since 1985, almost 75% are under 100 million barrels. A key issue for discoveries in that size range is proximity to infrastructure. Those fields in the deepest water furthest from existing pipelines may not be commercially developed until larger fields establish infrastructure nearby.

The large number of small discoveries may impact the direction of future environmental assessments as companies struggle to commercialize the deposits. Smaller fields will require sub-sea completions, longer production testing intervals to reduce reservoir performance uncertainty, and FPSO production schemes. Environmental impact of these issues will need to be understood in the discussion of commercial options.

While it is difficult to predict the geographic location of future environmental work, the drilling to date seems to be indicating two fairways of emerging geologic potential. One is located in the eastern Gulf, and extends from the southeastern portion of Green Canyon, through north-western Atwater Valley, and into south-eastern Mississippi Canyon. The large discoveries of Mad Dog, Atlantis, Thunder Horse and Tahiti are all located in this region. The second area is in the western Gulf, located in western Alaminos Canyon and eastern Keithly Canyon. This area, known as the Perdido fold-belt, is relatively unexplored, but the announcement of a recent discovery coupled with several clearly defined exploratory targets has focused industry interest in the area.

Workshop on Deepwater Environmental Studies Strategy: Industry Perspective
Click on the title to view the presentation.

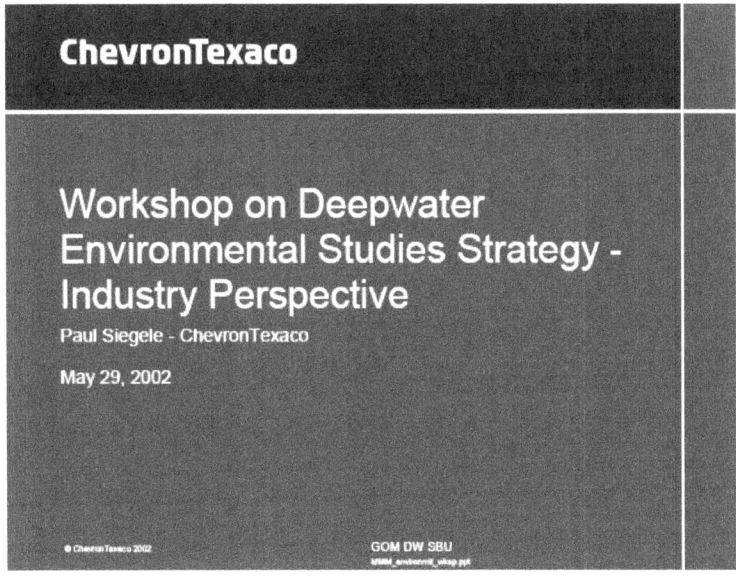

III. DeepSpill Experiment

Dan Allen
Chevron USA Production Co., New Orleans, Louisiana

Why should the DeepSpill Joint Industry Proposal (JIP) be conducted?

Little is known about the behavior of both liquid and gaseous hydrocarbon releases under deepwater conditions. Several questions were postulated that focused the JIP's efforts.

- Would hydrates be formed and how would they affect the continued release of the hydrocarbons?
- Would there be any gas phase dissolution? Would gas even come to the surface of the sea from a release?
- Would the released hydrocarbons eventually come to the sea surface and how far from the release site would this occur?
- Would phase separation occur?
- To what degree would entrainment of oil occur in the water column?
- What about the size distribution of the droplet and bubble from the hydrocarbon release?
- Would "weathering" occur on the liquid hydrocarbons in the water column?

The behavior of the released hydrocarbons could affect potential intervention methods. The formation of hydrates at the release point could limit or stop the release of the hydrocarbons. What kind of cleanup equipment might be needed, if it were needed at all?

Participants in the JIP wanted to optimize their spill response "toolbox." What would work for containment and cleanup of a deepwater spill? Could dispersants be used effectively? Tests were also needed on surveillance technologies to determine their applicability and effectiveness.

Deepwater computer transport models were being developed at two universities. Data from the deepwater release could be used to validate the model and its outcomes. Laboratory data were also generated from elements within the JIP. The release could serve to calibrate these data.

The DeepSpill JIP had some unlikely partners. The MMS contributed funding toward the JIP and was actively involved throughout the process. The Norwegian Pollution Control Authority (SFT) and the Norwegian Clean Sea Association (NOFO), a cleanup consortium, were participants in the JIP. Additionally, 22 oil companies rounded out the JIP partners.

Why was Norway selected for the test release? Norway has always been a leader in "field testing" oil release study efforts. The Norwegian environmental authority (SFT) and the area's oil spill response organization (NOFO) were involved in the study efforts. Since other releases have been successfully conducted off Norway, less legal complications were expected.

Conducting the Experimental Underwater Hydrocarbon Release

The table below depicts a summary of the four experimental discharges to be conducted during the study.

Table 1.
Summary of the four experimental discharges.

Experiment	Duration (minutes)	Gas Rate (Sm^3/s)	Water/Oil Rate (m^3/hr)
Nitrogen and dyed sea water	40	0.6	60
Marine diesel and LNG	60 (oil)	0.6	60
Crude oil and LNG	50 (oil)	0.7	60
LNG and sea water	120	0.7	60

A variety of instruments were used during the experimental release to monitor the characteristics and movement of the expelled fluids. Some instruments were selected to determine if they could accurately detect the released hydrocarbons within the water column. The following is a listing of some of the instruments used during the tests.

- Echosounders at 18, 38, 120, and 200 kHz
- Radar
- ADCP
- CTD, rosette sampler, PAH fluorimeter
- ROV videos

The following graphic shows a diagrammatic representation of the deployed equipment for the experimental hydrocarbon release.

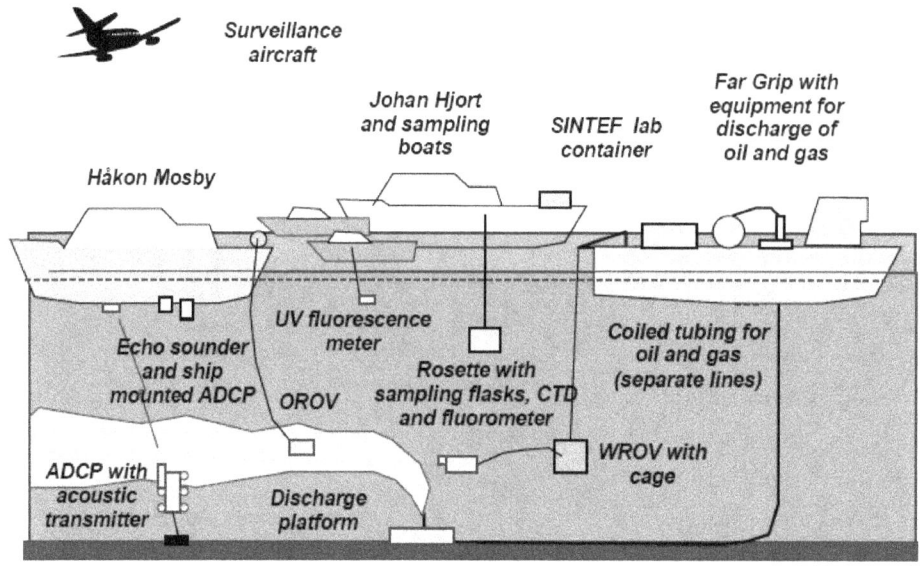

Figure 1. Diagrammatic of the deployed equipment for the experimental hydrocarbon release.

The experimental hydrocarbon release was to take place in the Norwegian sector of the North Sea at a location know as Heland Hansen. The map below shows this area and its relationship to the Norwegian coast.

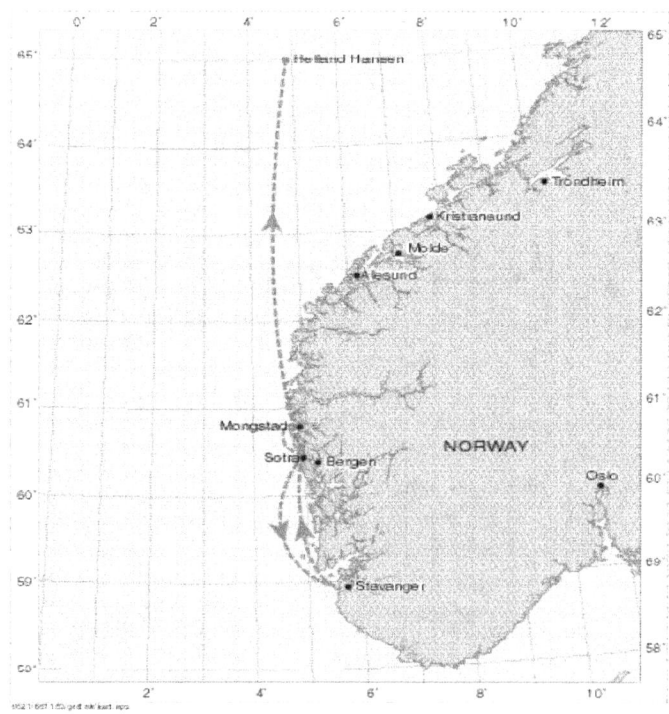

Figure 2. Relationship of experimental hydrocarbon release site to the Norwegian coastline.

The following picture shows the surface slick that developed during one of the experimental releases. Note the survey vessels working in the slick.

Figure 3. Surface slick that developed after a hydrocarbon test release.

The figure below shows an underwater gas release from the subsea manifold during one of the tests.

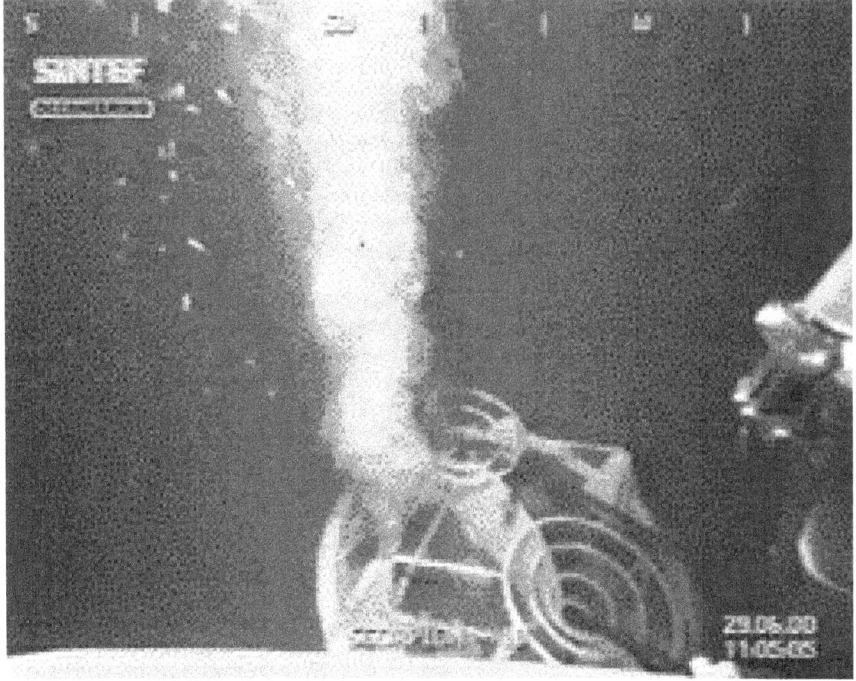

Figure 4. Underwater gas release from the subsea manifold.

Findings from the Experimental Deepwater Hydrocarbon Release.

The following list summarizes the major findings from the hydrocarbon release.

- While the temperature and pressure at the underwater release site was well within the ranges for stable hydrates to form, none were observed.
- The gas/oil plumes were clearly detected by echosounder equipment.
- No gas was observed at the sea surface.
- The oil surfaced near the underwater manifold site as thin films in about one hour after its release.
- Oil spill model results roughly agree with the observation at the test site.

DeepSpill
Click on the title to view the presentation.

IV. Technical Overviews

Overview of the Physical Oceanography of the Deep Gulf of Mexico

Ann E. Jochens, Steven F. DiMarco, Matthew K. Howard,
Worth D. Nowlin, Jr. and Robert O. Reid
Department of Oceanography, Texas A&M University, College Station, Texas

The deep Gulf of Mexico is the part of the Gulf that is seaward of the shelf break, which occurs approximately at the 200-m isobath. In an examination of the available current measurements in the deep Gulf, Nowlin et al. (2001) determined the background currents by extracting the maximum, mean, and standard deviation for each instrument. They found: that the highest maximum and mean current speeds were near the sea surface with maxima reaching up to 200 cm·s^{-1} in the eastern Gulf and 100 cm·s^{-1} in the western Gulf, that speeds decreased with depth tending toward a minimum near 1000 m, and that speeds increased somewhat with depth below that level, likely due to bottom intensification of currents. This suggests that the circulation and water properties within the deep Gulf can be viewed as being approximately a two-layer system in the vertical, divided approximately at the 1000-m depth.

There are two principal forcing functions of the circulation in the upper 800 to 1000 m of the deep Gulf of Mexico: the Loop Current system, including the Loop Current itself, Loop Current eddies (LCEs) and other circulation phenomena derived therefrom (Figure 1), and wind stress (Nowlin et al. 2001; see also Continental Shelf Associates, Inc. 2000). The Loop Current system begins with the Yucatan Current, which enters the Gulf through the Yucatan Channel and then separates from the Campeche Bank, becoming the Loop Current. The Yucatan Current is westward intensified, with speeds in the current core of approximately 200 cm·s^{-1} or more and dropping to about 50 cm·s^{-1} or less within 10 km to the west of the core and 100 km to the east (Cochrane 1963). There can be flows southward out of the Gulf at the western and eastern edges of Yucatan Channel (Cochrane 1963; Ochoa et al. 2001; Nowlin et al. 2001).

As it enters the Gulf the Yucatan Current extends to the sill depth of approximately 2000 m. Its continuation in the Gulf, the Loop Current, enters and exits the Gulf through the Yucatan Channel and the Florida Straits, respectively. It carries within it the water masses from the global ocean, originating mainly in the Atlantic Ocean (Morrison and Nowlin 1982). The sill depths of the entrance and exit are approximately 2000 and 800 m, respectively, and essentially control which water masses enter and leave the Gulf. Although buoyancy forcing by river discharge affects the nearshore coastal currents and is a form of thermohaline forcing known to be important over the shelves of the Gulf, no thermohaline forcing of consequence or water mass formation are known to occur in the deepwater region of the Gulf (Nowlin et al. 2000).

Water masses carried into the Gulf by the Loop Current include the Subtropical Underwater, which is characterized by a salinity maximum at typical depths of 150-250 m in the eastern Gulf (Morrison and Nowlin 1977) and 0-250 m in the western Gulf (Morrison et al. 1983). Below the Subtropical Underwater is 18°C Sargasso Sea water, which is characterized by an oxygen maximum of between 3.6-3.8 mL·L^{-1} at depths of 200-400 m in the Loop Current waters in the eastern Gulf (Morrison and Nowlin 1977). This water mass characteristic appears to erode away as the waters mix into the western Gulf. Another water mass is the Tropical Atlantic Central Water, which is characterized by an oxygen minimum of below 3 mL·L^{-1} that occurs at depths of 450-700 m in the eastern Gulf (Morrison and Nowlin 1977) and 250-500 m in the western Gulf (Morrison et al. 1983). The Antarctic Intermediate Water (AAIW) occurs between 700-1000 m depth in the eastern Gulf and 500-1100 m in the western Gulf and is characterized by nutrient maxima and salinity minimum. Below the AAIW is a modified upper North Atlantic Deep Water found between 900-1200 m in the eastern Gulf and 1000-1100 m in the western Gulf and characterized by a silicate maxima (Nowlin and McLellan 1967; Morrison and Nowlin 1977; Morrison et al. 1983).

Below the sill depth of the Yucatan Channel to the bottom, data suggest no clearly discernible horizontal variation in potential temperature, salinity, and dissolved oxygen distributions and only slight vertical gradients (McLellan and Nowlin 1963; Nowlin and McLellan 1967; Nowlin et al. 1969). The relatively uniform dissolved oxygen, potential temperature, and salinity of the waters below the sill depth indicate either that the deep waters have common sources or that the residence time is great enough to erode away gradients by exchange processes (Nowlin 1972). A recent modeling study, however, suggests relatively short residence times. Moreover, vertical variation (e.g., bottom intensification) of deep currents must be supported by density gradients.

The Loop Current extends into the eastern Gulf and then exits through the Florida Straits. As it extends into the Gulf, the Loop Current may be confined to the southeastern Gulf, as shown in the lower panel of Figure 1, or may extend well into the northeastern or north central Gulf, as shown in the upper panel of Figure 1 (see also Huh et al. 1981; Paluszkiewicz et al. 1983). Currents associated with the Loop Current and its eddies extend in depth at least down to the sill depth of the Florida Straits (approximately 800 m) and have surface speeds that may exceed 200 cm·s^{-1} and speeds of 10 cm·s^{-1} at 700 m (e.g., Hamilton 1997; Cooper et al. 1990; Molinari and Morrison 1988). Transport estimates of the Loop Current are approximately 25-30 Sverdrups (e.g., Sheinbaum et al. 2002; Ochoa et al. 2001; Maul and Vukovich 1993).

Anticyclonic Loop Current Eddies (LCEs) separate from the Loop Current at non-periodic intervals ranging from 3 to 17 months, with primary peaks in frequency of separation of 6 and 11 months and a secondary peak at 9 months (e.g., Sturges and Leben 2000; see also Sturges 1994). They have diameters that typically exceed 250 km when newly formed and that decrease with age as the LCE decays (Elliott 1982). The LCEs may remain in the eastern Gulf for some time after their formation and may even reattach to the Loop Current (e.g., Vukovich 1995). Most eventually move westward, with typical translation speeds of approximately 5 km·d^{-1} (Elliot 1982, see also Hamilton et al. 1999), and reach the western edge of the Gulf of Mexico basin. The LCEs have current cores and water properties similar to those of Caribbean type. Thus, their westward migration carries those water properties into the western Gulf. These anticyclonic, mesoscale current rings have average lifetimes longer than one year (Elliott 1982) and may spawn cyclonic rings during interaction with one another or with the continental slope (e.g., Vidal et al. 1994). Little is known of the velocity fields within the cyclones or small anticyclones (e.g., see Figure 1). Hamilton (1992) reported that slope cyclones in the central Gulf off Louisiana were long lived and had currents of 30 to 50 cm·s^{-1} in the upper layers and extended to depths of 200 to 800 m. Berger et al. (1996) reported on the existence of smaller anticyclones, but not on their velocity structures.

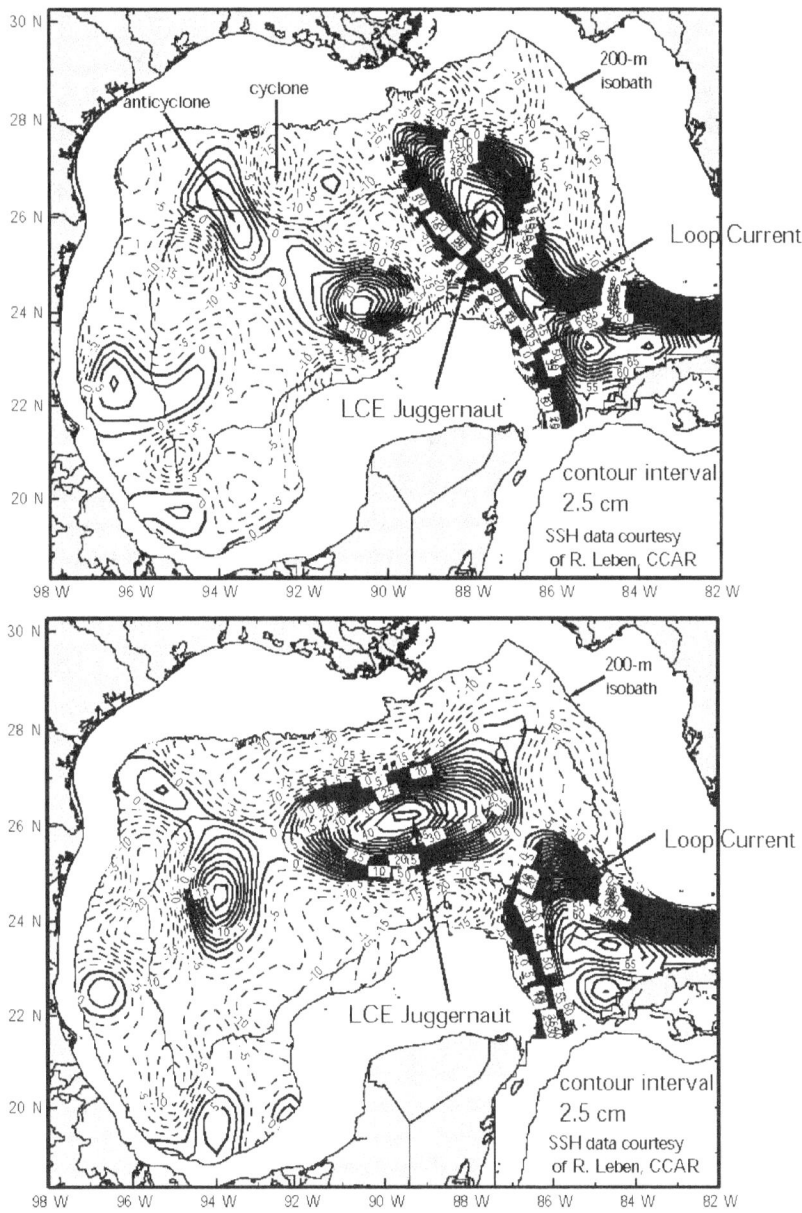

Figure 1.   Sea surface height field for 20 September 1999 (top) showing extension of Loop
            Current into the northeastern Gulf and formation of Loop Current Eddy Juggernaut and
            for 14 November 1999 (bottom) showing westward migration of LCE Juggernaut as well
            as the Loop Current in the southeastern Gulf and an older anticylone in the western Gulf.

The second forcing function is wind stress, which primarily affects the upper waters of the Gulf of Mexico. This forcing has two main classes of phenomena. First, low frequency regional wind patterns may generate low frequency regional circulation patterns. Sturges (1993) suggests a wind-driven anticyclonic circulation occurs over the west and central Gulf with a westward intensified boundary current adjacent to the western slope of the Gulf. It is centered north-south at approximately 24°N. The volume transport and speeds in this feature have an annual cycle, with maximum values in July and minimum in October, in response to the wind stress curl over the western Gulf. The transport varies from about 2.5 to 7.5 Sverdrups; the maximum drift speeds in the western boundary current vary from about 5 to 30 cm·s^{-1}. Vázquez de la Cerda (1993) has shown that seasonal variations in the wind fields over the Gulf generate a cyclonic circulation, centered near 20°N 94.5°W, over the Campeche Bay in the southwestern Gulf. This cyclone has an apparent annual signal with maximum transports of order 3 Sv in the summer, consistent with the minimum positive wind torque that occurs in that season.

The second class of wind forcing leads to the episodic current events that are forced by occasional atmospheric events such as tropical cyclones, extratropical cyclones, cold air outbreaks, and other frontal passages. The strongest of these includes hurricanes which can produce currents in the mixed layer that exceed 150 cm·s^{-1}, and when combined with wave currents, may exceed 300 cm·s^{-1} (Nowlin et al. 2000). Hurricane induced currents at 200 m have been observed to be approximately 100 cm·s^{-1} in several instances, and even at 700 m may reach 15-20 cm·s^{-1} (e.g., Brooks 1983). Tropical conditions normally prevail over the Gulf from May or June until October or November; the nominal hurricane season is 1 June through 30 November. From October or November until March or April the Gulf experiences intrusions of cold, dry continental air masses from the north that can result in cold air outbreaks and the formation of extratropical cyclones. These cyclogenesis events occur ~10 times per season. Although their main effects may be over the continental shelves, they can cause very energetic currents over the upper continental slopes. Speeds of 50-75 cm·s^{-1} in the surface layer and exceeding 20 cm·s^{-1} down to depths of 200 m have been observed.

Mention must be given to the occasional observations of subsurface-intensified currents of short duration, usually lasting of the order of a day, and with vertical extents of less than 100 m (Nowlin et al. 2001). High-speed current cores have been reported with maxima in the depth range of about 100 to 400 m and with maximum speeds of 100 cm·s^{-1} (or perhaps greater). Occurrences seem confined to regions over the upper continental slope. Their causal mechanisms are yet unknown.

Below about 800 m, the direction and speeds of deep currents are not well understood. Model output and geostrophic calculations based on reasonable choice of reference levels suggest the long-term mean circulation of the deep Gulf is cyclonic (Nowlin et al. 2001). This cyclonic circulation also seems to be intensified offshore of the steep topography of the Sigsbee Escarpment in the north central Gulf, the Campeche shelf, and the west Florida shelf.

Using a sparse array of current meters and a numerical model, Hamilton (1990) and Sturges et al. (1993), respectively, suggested that currents below 1000 m may be strongly influenced by the Loop Current through excitation of energetic currents associated with topographic Rossby waves. Hamilton (1990) inferred that low-frequency fluctuations with periods greater than 10 days propagated from the eastern to western Gulf with group speeds near 9 km·d^{-1}. Current speeds within these deep motions have been observed to be approximately 30 cm·s^{-1} in the domain beneath the Loop Current, but less than 20 cm·s^{-1} in the central and western Gulf. These currents were barotropic, being highly coherent in the vertical, and exhibited bottom intensification (Figure 2). Evidence suggests these deep currents may be excited by the Loop Current or LCE separations.

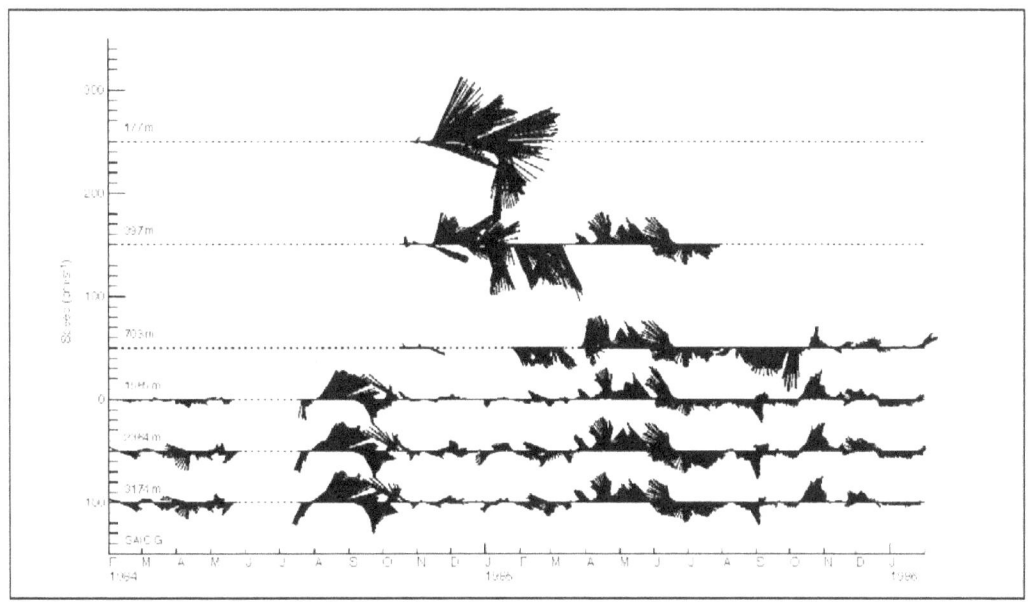

Figure 2. Current vectors (40-hr low-passed; north directed upward) from SAIC mooring G located in 3,200 m water depth at 25.60 N, 85.50 W off the southern West Florida Shelf. In December 1984, currents above 1,000 m are affected by the Loop Current in a surface-intensified manner, while currents below are not affected. During April-June 1985, during an eddy separation, the currents appear to be coherent throughout the water column, with some indication of bottom intensification.

Additional information on the currents beneath 800-1000 m comes from numerical models. Model studies suggest that both cyclonic and anticyclonic eddies are present, perhaps as pairs, in the deep basin waters (Sturges et al. 1993; Welsh and Inoue 2000; Nowlin et al. 2001). The eddies form in the eastern Gulf under the Loop Current or newly formed LCEs. Guided by topography, they then propagate into the western Gulf. Some model results suggest deep layer eddies form, often as a pair consisting of an anticyclone and a stronger cyclone, in a response to a westward moving newly formed anticyclonic LCE in the upper layer. Evidence from one model suggests the deep eddy pair stays coupled with the upper LCE as it migrates westward, with the two deep rings rotating cyclonically as a pair, and with the deep anticyclone decaying more rapidly than the deep cyclone (Welsh and Inoue 2000; see also Nowlin et al. 2001).

Finally, intense bottom currents may be responsible for a bed form, consisting of mega-furrows with depths of 5 to 10 m and widths of 10s of m eroded into Holocene deposits, discovered by William Bryant of Texas A&M University in early 1999 on the continental slope and rise of the north-central Gulf of Mexico. These furrows are spaced on the order of 100 m apart and extend unbroken for distances of order 100 km, generally oriented nearly along isobaths. Water depths range from 2000 to 3000 m. It seems likely that the processes responsible for these furrows are active at present. Speculation based on laboratory experiments is that near-bottom speeds of currents responsible for the inshore furrows might be 50 $cm \cdot s^{-1}$ or even in excess of 100 $cm \cdot s^{-1}$. Nearly barotropic currents fluctuating essentially along-isobath with amplitudes reaching 100 $cm \cdot s^{-1}$ and periods of order 10 d have been observed above the bottom in the region just inshore from the mega-furrows. The currents associated with furrows might be sporadic or quasi-permanent. They may be related to the mean circulation within the deep basin or possibly to topographic waves or deep eddies.

References

Berger, T. J., P. Hamilton, J. J. Singer, R. R. Leben, G. H. Born and C. A. Fox. 1996. Louisiana/Texas shelf physical oceanography program: eddy circulation study, final synthesis report. Volume I: Technical Report. OCS Study MMS 96-0051. U.S. Dept. of the Interior, Minerals Management Service, Gulf of Mexico OCS Region, New Orleans, LA. 324 pp.

Brooks, D. A. 1983. The wake of Hurricane Allen in the western Gulf of Mexico. J. Phys. Oceanogr. 13:117–129.

Cochrane, J. D. 1963. Yucatan Current. Pp. 6-11, In Oceanography and Meteorology of the Gulf of Mexico, Annual Report, 1 May 1962 - 30 April 1963. Texas A&M University, Department of Oceanography, Tech. Rpt. Ref. 63-18A, College Station, TX.

Continental Shelf Associates, Inc. 2000. Deepwater Gulf of Mexico environmental and socioeconomic data search and literature synthesis. Volume I: Narrative Report, OCS Study MMS 2000-049. Volume II: Annotated bibliography, OCS Study MMS 2000-050. U.S. Dept. of the Interior, Minerals Management Service, Gulf of Mexico OCS Region, New Orleans.

Cooper, C., G. Z. Forristall and T. M. Joyce. 1990. Velocity and hydrographic structure of two Gulf of Mexico warm-core rings. J. Geophys. Res. 95(C2):1663-1679.

Elliott, B. A. 1982. Anticyclonic rings in the Gulf of Mexico. J. Phys. Oceanogr. 12(1):292–1,309.

Hamilton, P. 1990. Deep currents in the Gulf of Mexico. J. Phys. Oceangr. 20:1087-1104.

Hamilton, P. 1992. Lower continental slope cyclonic eddies in the central Gulf of Mexico. J. Geophys. Res. 97(C2):2185-2200.

Hamilton, P. 1997. Chapter 2 –The Physical Environment. In Northeastern Gulf of Mexico Coastal and Marine Ecosystem Program: Data Search and Synthesis; Synthesis Report. Science Applications International Corporation. U.S. Department of the Interior, U.S. Geological Survey, Biological Resources Division, USGS/BRD/CR-1997-0005 and Minerals Management Service, Gulf of Mexico OCS Region, New Orleans, LA, OCS Study MMS 96-0014.

Hamilton, P, G. S. Fargion and D. C. Biggs. 1999. Loop Current eddy paths in the western Gulf of Mexico. J. Geophys. Res. 29(6):1180-1207.

Huh, O. K., W. J. Wiseman, Jr. and L. J. Rouse, Jr. 1981. Intrusion of Loop Current waters onto the west Florida continental shelf. J. Geophys. Res. 86(C5):4186–4192.

Maul, G. A. and F. M. Vukovich. 1993. The relationship between variations in the Gulf of Mexico Loop Current and Straits of Florida volume transport. J. Phys. Oceangr. 23:785-796.

McLellan, H. J. and W. D. Nowlin, Jr. 1963. Some features of the deep water in the Gulf of Mexico. J. Mar. Res. 21:233–245.

Molinari, R. L. and J. Morrison. 1988. The separation of the Yucatan Current from the Campeche Bank and the intrusion of the Loop Current into the Gulf of Mexico. J. Geophys. Res. 93(C9):10645-10654.

Morrison, J. M., and W. D. Nowlin, Jr. 1977. Repeated nutrient, oxygen, and density sections through the Loop Current. J. Mar. Res. 35(1):105–128.

Morrison, J. M. and W. D. Nowlin, Jr. 1982. General distribution of water masses within the eastern Caribbean Sea during the winter of 1972 and fall of 1973. J. Geophys. Res. 87(C6):4207-4229.

Morrison, J. M., W. J. Merrell, Jr., R. M. Key and T. C. Key. 1983. Property distributions and deep chemical measurements within the western Gulf of Mexico. J. Geophys. Res. 88(C4):2601-2608.

Nowlin, W. D., Jr. 1972. Winter Circulation Patterns and Property Distributions. Pp. 3–5 in: L. R. A. Capurro and J. L. Reid (Eds.), Contributions on the Physical Oceanography of the Gulf of Mexico. Texas A&M University Oceanographic Studies, Volume 2,1. Gulf Publishing Co., Houston, 288 pp.

Nowlin, W. D., Jr., A. E. Jochens, S. F. DiMarco, R. O. Reid and M. K. Howard. 2001. Deepwater Physical Oceanography Reanalysis and Synthesis of Historical Data: Synthesis Report. OCS Study MMS 2001-064, U.S. Dept. of the Interior, Minerals Management Service, Gulf of Mexico OCS Region, New Orleans, LA. 528 pp.

Nowlin, W. D., Jr., A. E. Jochens, S. F. DiMarco and R. O. Reid. 2000. Chapter 4: Physical Oceanography. Pp. 61-121, In Continental Shelf Associates, Inc., 2000.

Nowlin, W. D., Jr., D. F. Paskausky and H. J. McLellan. 1969. Recent dissolved-oxygen measurements in the Gulf of Mexico deep waters. J. Mar. Res. 27(1):39-44.

Nowlin, W. D., Jr., and H. J. McLellan. 1967. A characterization of the Gulf of Mexico waters in winter. J. Mar. Res. 25:25–59.

Ochoa J., J. Sheinbaum, A. Badan, J. Candela and D. Wilson. 2001. Geostrophy via potential vorticity inversion in the Yucatan Channel. J. Marine Res. 59(5):725-747.

Paluszkiewicz, T., L. A. Atkinson, E. S. Posmentier and C. R. McClain. 1983. Observations of a Loop Current frontal eddy intrusion onto the west Florida shelf. J. Geophys. Res. 88(C14):9639–9651.

Sheinbaum, J., J. Candela, A. Badan and J. Ochoa. 2002. Flow structure and transport in the Yucatan Channel. Geophys. Res. Letters 29(3):10-1 to 10-4.

Sturges, W. 1993. The annual cycle of the western boundary current in the Gulf of Mexico. J. Geophys. Res. 98:18,053–18,068.

Sturges, W. 1994. The frequency of ring separations from the Loop Current. J. Phys. Oceanogr. 24:1,647–1,651.

Sturges, W. and R. Leben. 2000. Frequency of ring separations from the Loop Current in the Gulf of Mexico: a revised estimate. J. Phys. Oceanogr. 30:1,814–1,819.

Sturges, W., J. Evans, S. Welsh and W. Holland. 1993. Separation of warm-core rings in the Gulf of Mexico. J. Phys. Oceanogr. 23:250–268.

Vázquez De la Cerda, A. M. 1993. Bay of Campeche Cyclone. Ph.D. Dissertation, Texas A&M University, 91 pp.

Vidal, V. M. V, F. V. Vidal, A. F. Hernandez, E. Meza and J.M. Perez-Molero. 1994. Baroclinic flows, transports, and kinematic properties in a cyclonic-anticyclonic-cyclonic ring triad in the Gulf of Mexico. J. Geophys. Res. 99(C4):7571-7597.

Vukovich, F. M. 1995. An updated evaluation of the Loop Current's eddy-shedding frequency. J. Geophys. Res. 100(C5):8655-8659.

Welsh, S. E. and M. Inoue. 2000. Loop Current rings and the deep circulation in the Gulf of Mexico. J. Geophys. Res. 105:16,951–16,959.

Overview of the Physical Oceanography of the Deep Gulf of Mexico

Ann E. Jochens
Texas A&M University

MMS Workshop on Deepwater Environmental Studies Strategy: A Five-Year
Follow-up and Planning for the Future, May 29-31, 2002

Overview of the Geology/Geohazards of the Continental Slope of the Northwest Gulf of Mexico

William R. Bryant
Department of Oceanography, Texas A&M University, College Station, TX

The engineering and geological constraints on the continental slope off Texas and Louisiana related to hydrocarbon recovery will require both novel geological and geophysical surveys and engineering methods. Significant seafloor engineering problems in deep water include slope instabilities, both short-term (slump) and long-term (creep); pipeline spanning problems; mass transport from unknown causes; and unusual stiffness and strength conditions. The geohazards (engineering and geologic constraints) present in and on the central and western continental slope are many in number and are mainly due to the activity of salt and rapid sedimentation. Specific examples include: faults (sediment tectonics, halokinesis); slope stability (slope steepening, slumps, creep, debris flow); gassy sediments (sediment strength reduction, hydrates, sediment liquefaction); fluid and gas expulsion features; diapiric structures (salt, mud, hydrates); seafloor depressions (blowouts, pockmarks, seeps); seafloor features (sediment waves, differential channel fill, brine-low channels, seabed furrows); shallow water flow; and deep water high-velocity currents (mega-furrows, seabed erosion).

The Gulf of Mexico is unique in the construction and evolution of its northwestern continental margin and in particular the continental slope off Texas and Louisiana. The processes that determined the physiography of the continental slope are almost completely dominated by the halokinesis of allochthonous salt. Harry Roberts once noted that, "twenty-five years ago our knowledge of processes on the continental slope in the northern Gulf of Mexico was so limited that it was viewed simply as an accreting sedimentary structure." In contrast, the continental slope off Texas and Louisiana is now viewed as one of the most complex and dynamic passive continental margins in the world.

Bathymetric maps reveal the presence of over 90 intraslope basins with relief in excess of 150 m. Intraslope-interlobal and intraslope-supralobal basins occupy the upper/middle and lower continental slope, respectively. In addition to the many structural effects, salt movement creates subsurface conduits that allow liquids and gases to seep into near-surface sediments and the overlying water column. The extent of seepage on the slope as compared with the shelf is important from the standpoint of both chemistry and the chemosynthetic communities associated with seeps.

The near-surface geology and topography (the area of most concern in relationship to submarine slope stability) of the continental slope off Texas and Louisiana are a function of the interplay between episodes of rapid shelf edge progradation and contemporaneous modification of the depositional sequence by diapirism and mass movement processes. Many slope sediments have been uplifted, folded, fractured, and faulted by diapiric action. Over steepening on the basin flanks and resulting mass movements have resulted in the appearance of highly overconsolidated sediments underlying extremely weak pelagic sediments. The construction of the Mississippi Canyon is in part a function of sidewall slumping and pelagic drape of low shear strength sediments. In contrast, slope over steepening and subsequent mass movement have resulted in high pore pressures in rapidly deposited debris flows on the upper slope and on basin floors, resulting in unexpected decreased shear strengths. Biogenic and thermogenic gas induces the accumulation of hydrates and underconsolidated gassy sediments, which are common on the upper continental slope. On the middle and lower continental slope, gassy sediments are not common except in the basins that do not have a salt base such as Beaumont Basin. The salt nappe restricts the upward movement of gas from below. Holocene and Pleistocene sediment cores recovered from the continental slope off Texas and Louisiana from conventional piston coring and from DSDP activities reveal the presence of unconsolidated gassy clays, silty clays, sands, and clayey sands, many containing gas hydrates.

The intraslope intralobal basins located on the upper slope continental range in water depths from 1,500 to 2,200 m. The bathymetry of the Central and Western Gulf areas is shown in Figure 1. The bathymetry of the upper to middle continental slope area consists of relatively flat ridges and basin floors separated by intraslope escarpments. The intraslope basin escarpments have relief up to 700 m and slopes between 5° to 30° and in specific locations up to 50°. Ridges that rim the basins correspond to late laterally spreading flat-topped salt tongues overlain by a thin sediment cover (Bryant et al. 1995). The deeper portions of intraslope intralobal basins are salt free and exhibit a dissected topography consisting of a multitude of small submarine canyons along the walls. Cores taken on the wall of some basins indicate that as much as 3 m to 5 m of sediment has been removed by slumping. The intraslope supralobal basin on the lower continental slope where the physiography is comparatively smooth (Figure 1) shows that the relief exists mainly as a rounded depression. The formation of basins on the lower slope is where subsidence is accomplished by evacuation of underlying salt (salt withdrawal).

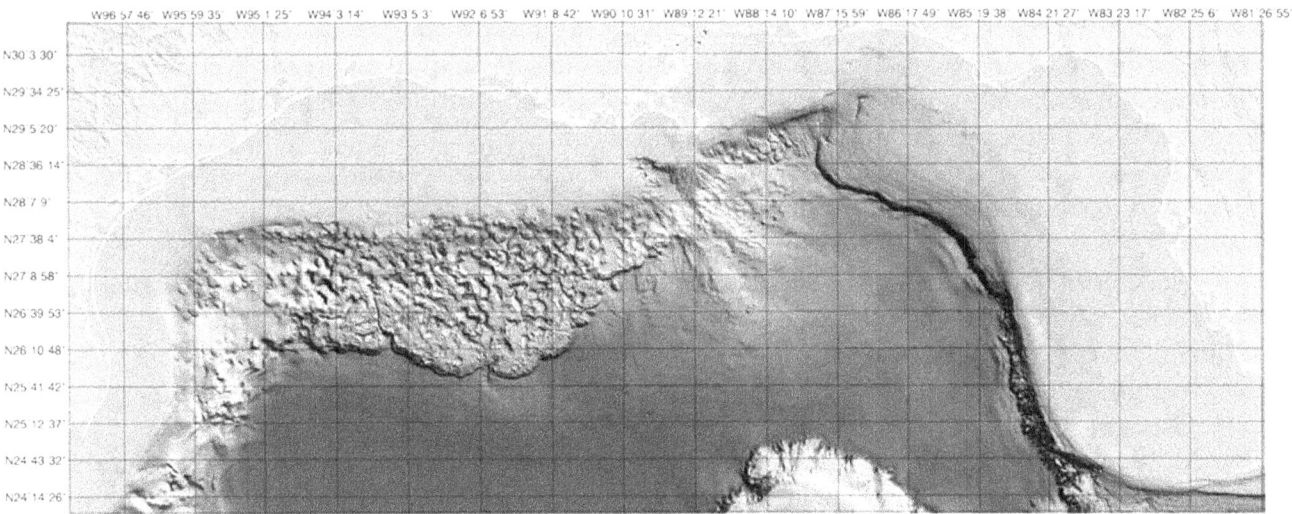

Figure 1. Bathymetry map of the northern Gulf of Mexico.

The submarine canyons along the Sigsbee Escarpment, Alaminos, Keathley, Bryant, Cortez, Farnella and Green Canyon are the result of the coalescing of salt canopies and nappes, the migration of the salt over the abyssal plain and the erosion of the escarpment during periods of low sea stand (Bryant et al. 1992). The bathymetry of the canyons is illustrated in Figure 1. In addition to the canyons that form along the escarpment, numerous small submarine canyons and gullies line the escarpment along with large slumps. Seaward of the canyons submarine fans of various sizes extend out onto the continental rise. A significant portion of the canyon walls and the escarpment contains slopes of 5° to 10° and slopes in excess of 15° are not rare. Large slope failures are present in the Green Canyon area especially in the Mad Dog, Neptune and Atlantis prospects.

The major faults on the continental slope are associated with massive accumulation of sediments and are called growth faults. These growth faults form contemporaneously and continuously with sediment deposition. The growth faults are found mostly on the upper continental slope and on the continental shelf where sediment accumulation is the thickest. The most common types of fault on the middle and lower continental slope are "groups of geometrically classified fault families and fault welds that are kinematically and genetically linked to each other and to associated salt bodies and welds. Linked fault systems can contain extensional, contractional, and strike-slip components. Extensional fault families are formed by basinward translation, subsidence into salt, or folding. Those fault families that accommodate basinward translation are balanced by salt extrusion or contract ional fault families" (Rowan et al. 1999). Rowan et al. related five associations of linked fault systems that are directly related to five types of salt systems: autochthonous salt (salt in place), stepped counterregional, roho, salt-stock canopy, and salt nappe. Faulting resulting from the formation of salt diapirs from autochthonous salt is the most common type fault on the upper slope while faulting from salt-stock canopy and salt nappe are most common on the middle and lower continental slope. Extensive faulting can be found on the rim of most intraslope intralobal and supralobal basin on the middle and lower continental slope. The faults are extensional faults caused by the upward movement of salt resulting from pressures created by sediment accumulation within basins. This type of faulting results in the occurrence of a large number of small faults in the area of the seafloor under going extension. In some areas of the slope the upward migration of salt results in the seafloor being totally fractured (faulted) and continuously displaced.

Portions of some of the submarine canyons, like Bryant Canyon, are being filled with salt due to the loading of the salt by sediments on the margins of the canyon. The salt migrates upward; filling the canyon that was created by turbidity current flow active during times of low-sea stand. The migration of salt into the canyon may occur at the rate of centimeters per year.

On the middle and lower continental slope, salt may be very close to the seafloor in certain areas and, on features such as the salt plug called "Green Knoll," salt is exposed at the seafloor and is being dissolved by seawater, resulting in the collapse of the cap of the knoll (Figure 2). In Orca Basin, an intraslope intralobal basin, salt is exposed at the bottom of the northern portion of the basin and a famous brine pool has formed within the basin. Brine flows imitating from exposed salt along the Sigsbee Escarpment have been observed in the Mad Dog prospect, northwest of Green Knoll. When salt is close to the seafloor, the emplacement of structures that require foundation pile new engineering methods will be necessary.

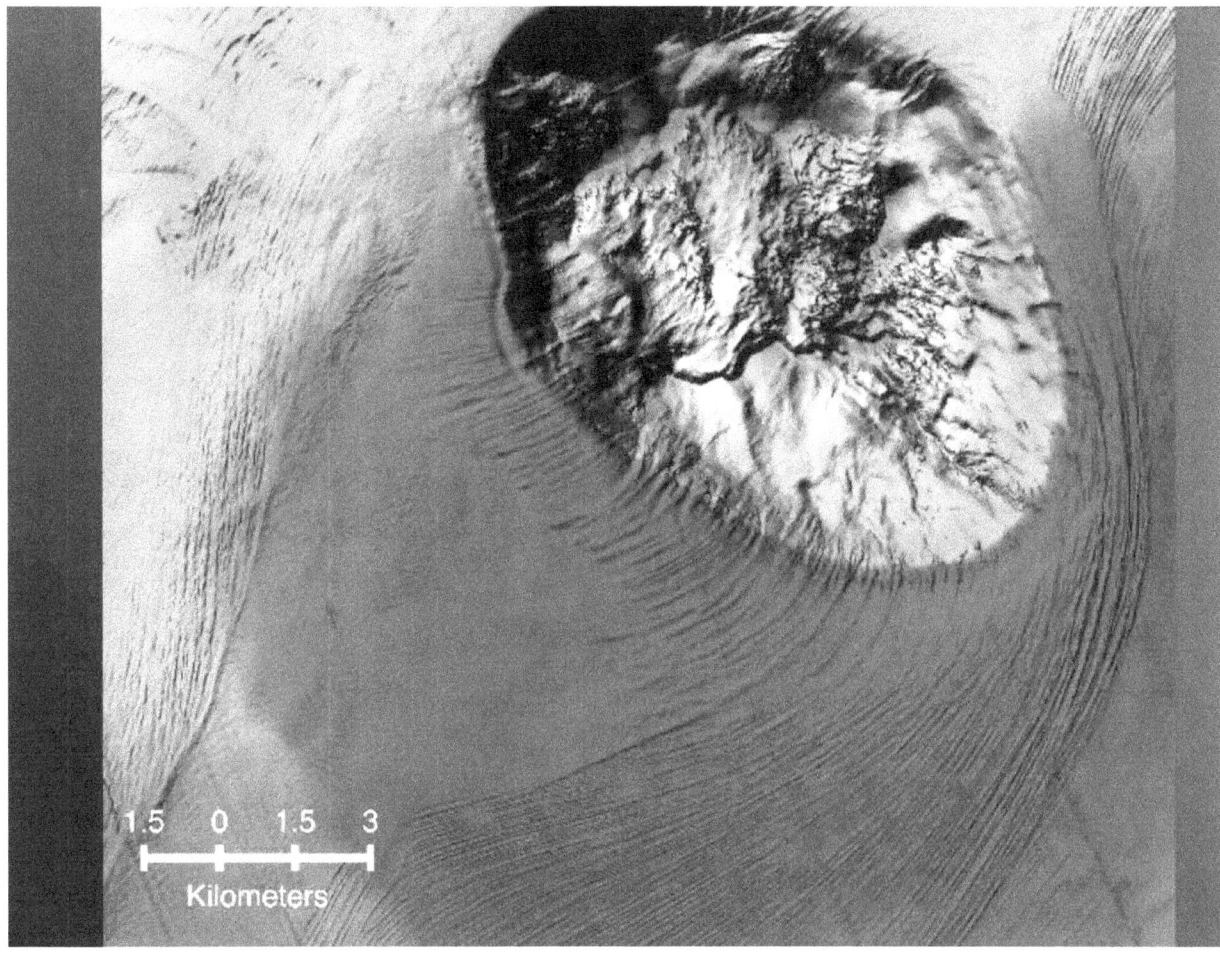

Figure 2. 3D surface extraction showing the physiography of Green Knoll. The collapsed top of the knoll is the result of seawater leaching the knoll's salt core. (WesternGeco 3D data).

Water currents can be a problem to structures on the continental slope, but they may be a major problem to structures such as platforms, bottom assemblies and pipelines at the base of the Sigsbee Escarpment starting in water depths as shallow as 1,200 m and as deep as 3,300 m. Recent studies have revealed the presence of large mega-furrows all along the base of the Sigsbee Escarpment. These large bedforms, 20 to 30 m wide and as deep as 10 m, occur along the base of the Sigsbee Escarpment and extend to a distance of 30 km seaward of the escarpment (Figure 3). The mega-furrows are the result of high velocity bottom currents occurring along the base of the Sigsbee Escarpment. The mega-furrows have been found at the base of the escarpment extending from 90° to 94.5° W Longitudes, a distance of over 500 km (Figure 4).

Figure 3. Mego-furrows on the continental rise South of Green Knoll. (WesternGeco 3D data).

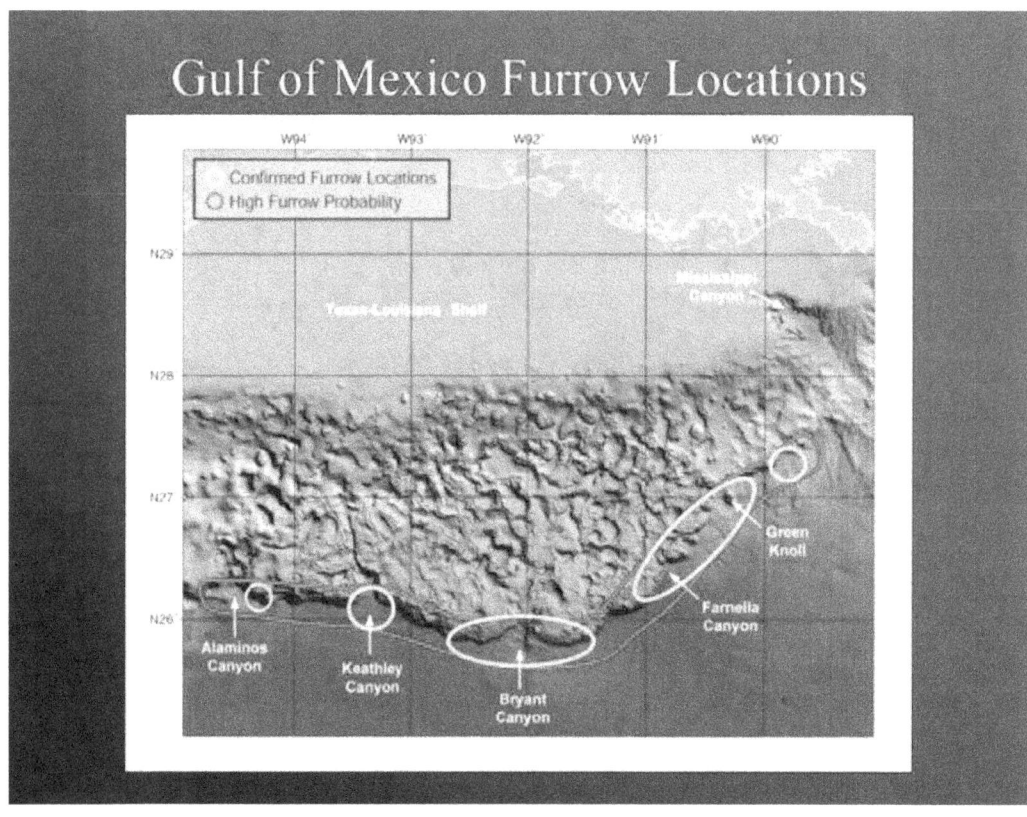

Figure 4. Map illustrating the location of mega-furrows along the continental rise seaward of the Sigsbee Escarpment, Northwest Gulf of Mexico.

Shallow water flow, also known as geopressured sands, is the uncontrolled flow of sand and water that can create significant sediment pile up at the wellhead. Shallow water flow is the result of excess pore pressures in the confining strata as the result of underconsolidation, also called compaction disequilibrium and differential compaction of fine-grained sediments. The flow usually occurs at 360 to 530 m below the seafloor. It is more likely to occur on the upper and middle continental slope and less likely to occur above the salt nappe, the tabular salt blocking the escape of overpressures from below.

Table 1 summarizes the properties related to geohazards of upper, middle, and lower continental slopes intraslope basins and lower slope canyons and the Sigsbee Escarpment.

Table 1.
Engineering constraints and possible geohazards of intraslope basins, canyons and escarpments.

Upper to Middle Slope Intraslope Interlobal Basins
- Steep sidewalls average 10° to 20°, maximum 50°
- Small submarine canyons and gullies dissect basin escarpments
- Basin wall sediments may be unstable and undergoing modification by creep and slump processes
- Low shear strength debris flow sediments on basin floor
- Basin floor subject to debris flows from side wall slumping
- Stiff sediments on highly faulted ridges between basins
- Hydrates, gas sweeps, carbonate bioherms and chemosynthetic organisms may be present
- Basins may contain low shear strength gassy anoxic sediments
- Isolated basins subject to formation of brine pools
- Basin sediments underconsolidated at shallow subbottom depths

Lower Slope Intraslope-Supralobal Basins
- Elevated faulted ridges between basins
- Elevated ridge along basin rim
- Basins are bowl shaped with low angle basin floor
- Soft surficial sediments within basin
- Structures on basin floor subject to debris flow
- Basin sediments underconsolidated at shallow subbottom depths

Lower Slope Canyons and Escarpments
- Side walls average 10° to 15°, maximum 45°
- Small submarine canyons and gullies dissect escarpment and smaller canyon escarpments
- Canyons and escarpment structurally active from effects of halokinesis
- Very rugged topography
- Slump deposits and slope failure common
- Small submarine fans on canyon floor formed from debris flows and turbidity currents
- In very deep water
- Sediments underconsolidated at shallow subbottom depths
- High velocity bottom currents and mega-furrows present at base of Sigsbee Escarpment
- Density currents along the base of the escarpment resulting from the leaching of exposed salt along the escarpment.

References

Bryant, W. R., G. S. Simmons and P. Grim. 1992. The morphology and evolution of basins on the continental slope northwestern Gulf of Mexico. Gulf Coast Association of Geological Societies Transactions 41:73-82.

Bryant, W. R., J. Y. Liu and J. Ponthier. 1995. The engineering and geological constraints of intraslope basins and submarine canyons of the northwestern Gulf of Mexico. Gulf Coast Association of Geological Societies Transactions 45:95-101.

Rowan, M. G., P. A. Jackson and B. D. Trudgill. 1999. Salt-related fault families and fault welds in the northern Gulf of Mexico. AAPG Bulletin 83(9):1,454-1,482.

Presentation Not Available

Overview of Knowledge of the Benthos of the Deep Gulf of Mexico

Gilbert T. Rowe
Department of Oceanography, Texas A&M University, College Station Texas

Introduction

The exploration of the biota of the deep Gulf of Mexico (GOM) can be divided historically into more or less three eras. The earliest explorations were concerned mostly with dredging the deep-sea floor. The most extensive work is summarized by Agassiz (1888). Few generalizations could be made in these works; they dealt mostly with lists of species and descriptions of new species. Publications resulted on a broad suite of invertebrate taxa. None of the early studies penetrated the western Gulf. These early investigations have been reviewed by Galtsoff (1954) and Geyer (1970).

The second important period of deep Gulf studies was initiated by Willis E. Pequegnat in the mid-1960's. A diverse series of investigations employed the *R/V Alaminos*. Efforts were made to characterize all geographic areas of the deep Gulf and the far western Caribbean. The composite samples were divided up by taxon or by specialized ecological problems among the graduate students at Texas A&M working with Pequegnat. The individual studies are available at Texas A&M or at University Microfilms. Many sampling trips extended well into what is now recognized as Mexico's Exclusive Economic Zone (EEZ) because at the time no restrictions had been placed on the scientific sampling of these international waters. A similar geographic distribution of studies would not be possible today without permission from Mexico. These studies have been summarized by Pequegnat (1983) in a volume submitted to the Minerals Management Service of the U.S. Dept. of the Interior. The work concentrated on samples taken with a sampling device developed by Pequegnat called a "skimmer" (Pequegnat et al. 1970). It was equipped with an odometer wheel to estimate distance covered. The anterior "mouth" of the skimmer measured 3 m wide by 1 m high. The anterior frame of the rigid, hour glass-shaped structure was covered with 1.25 cm galvanized wire mesh but its bulbous cod end was covered with 0.6 cm mesh. The distances traveled averaged several kilometers. The original meter wheel data are available in the field notes taken aboard ship at the time of sampling. The significant thing about this method is that it sampled "megafauna," principally, and not other size categories. It had the advantage that the samples were protected by the rigid posterior framework, and thus fragile forms were damaged less than in conventional otter or beam trawls.

Faunal groups studied from the deep Gulf included the crustaceans (Pequegnat 1970; Roberts 1970; Firth 1971), echinoderms (Carney 1971; Booker 1971), molluscs (James 1972), and fishes (Bright 1968), among others. Tabulation of fish gut contents were intended to link the megafauna to its food source (Rayburn 1975). Kennedy (1976) compared the species composition of the eastern and western Gulf faunas and concluded that they were not the same. The macrofauna appeared to be grouped into assemblages that were distributed within zones down the slope onto the abyssal plain (Kennedy 1976; Roberts 1977), but no justification was found for separating the abyssal plain fauna into zoogeographic provinces by latitude or longitude. That is, even though the fauna along the slope changed with increasing depth, and that these changing faunas differed in the east and west Gulf, down on the basin floor the fauna showed little in the way of east - west differences.

The stations occupied by the *Alaminos* also incorporated bottom multi-shot photography to quantify areas of the sea floor. The principal rationale for this was to quantify the densities of the epibenthic megafauna captured by the skimmer and other miscellaneous trawls and dredge samples. It also allowed the users to count "lebenspuren" (animal tracks and burrows) and qualitatively categorize the types of sediments present. Although a 70 mm format camera was used on occasion, a 35 mm camera built by Alpine Geophysical seems to have produced more good film. It was rigged to take photographs on bottom contact of a trigger switch. The advantage of such an approach is that distance between the camera and the bottom is always known and this allows the operator to know the area each photograph covers with a given lens angle. This can be calibrated with a grid in shallow water, if necessary. The records available indicate that on the order of up to 50 photographs were taken per lowering and each photograph covered an area of ca. 1 m^2. An appendix in Pequegnat (1983) features a large number of the photographs.

In addition to the biological studies, photographs of an "iron stone bottom" north of the Yucatan Strait suggested deep bottom currents can be strong enough in the eastern Gulf to sweep large areas free of sediments (Pequegnat 1970). The negatives of the entire collection of lowerings are archived in the Oceanography Department at Texas A&M University (with the present exception of a station on loan to the Universidad Nationale Autonoma de Mexico).

Abyssal plains constitute a significant proportion of the surface of the earth. Although generally acknowledged as the ultimate sink of detritus from the continents, their biota has long been considered sparse and depauperate. The abysso-benthic communities of the Sigsbee Deep in the western Gulf of Mexico, known as one of the flattest surfaces on the deep ocean floor, is less well known than many other abyssal plains, in spite of its modest depth (3.6 to 3.8 km) and proximity to the United States and Mexico. The works of Pequegnat remain the most extensive on Sigsbee Deep plain.

The third historical time-frame represents the most extensive sampling of the sediment biota of the continental slope of the northern Gulf of Mexico to date. This was conducted in the 1980's by LGL Environmental, with support from the MMS. This consisted of paired Gray-O'Hara or GOMEX box cores (Boland and Rowe 1991), bottom survey camera lowerings and bottom trawling. The stations studied included three transects down the continental slope off Texas, Louisiana and Florida (Figure 1). Sampling stopped at depths just shy of 3 km and therefore did not extend out onto the abyssal plain. Thus, comparison with shelf and abyssal plain samples was not possible.

The work on deep-water benthos at Texas A&M beginning in the 1960's up through the MMS-supported studies of the 1980's has been published in a concise summary by Pequegnat et al. (1990). The focus was on the northern continental slope. Other documentation of the studies of the slope includes reports to MMS (Gallaway 1988; Gallaway et al. 1988) and a dissertation on the polychaete annelid worms (Hubbard 1995).

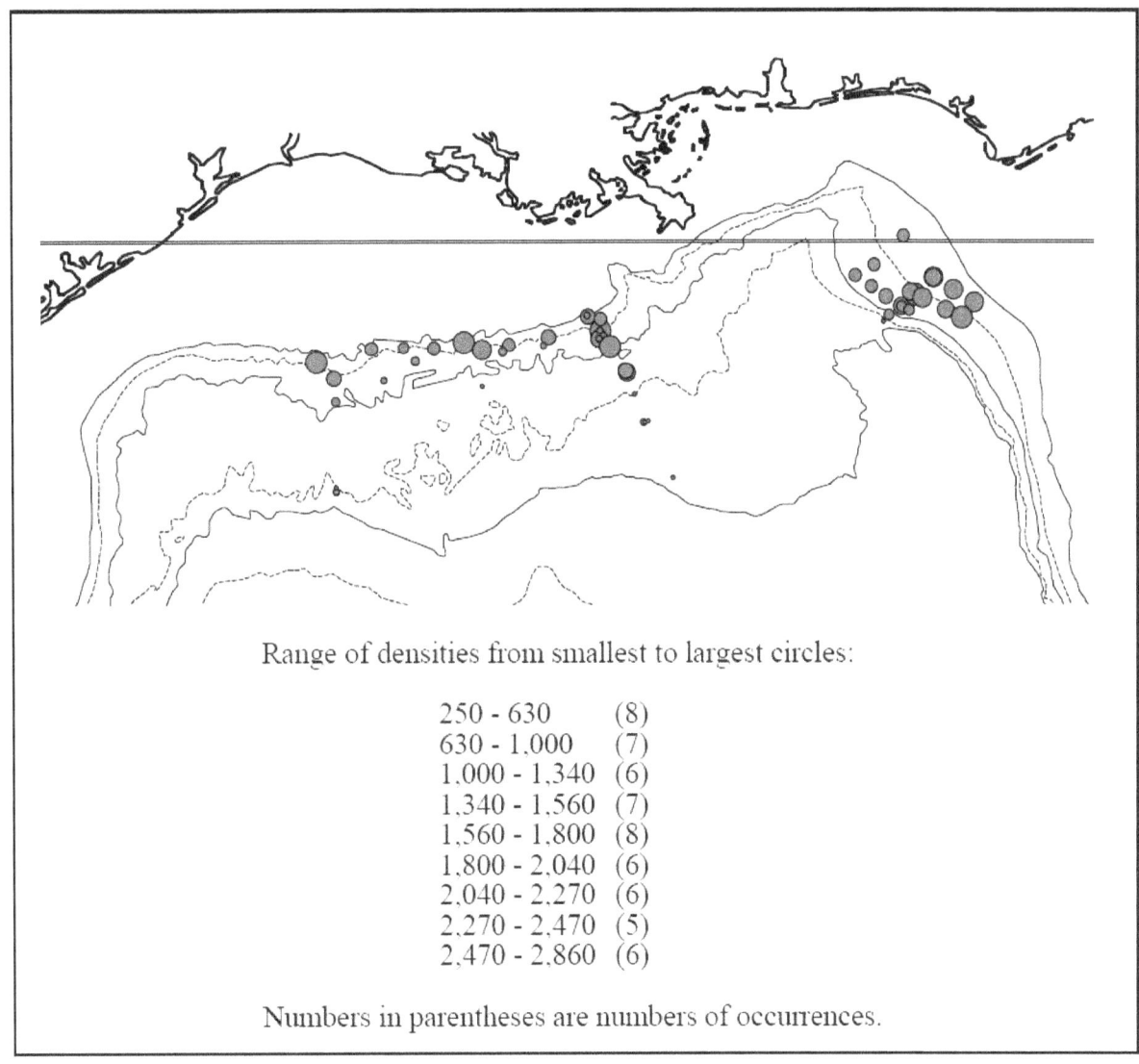

Range of densities from smallest to largest circles:

250 - 630	(8)
630 - 1,000	(7)
1,000 - 1,340	(6)
1,340 - 1,560	(7)
1,560 - 1,800	(8)
1,800 - 2,040	(6)
2,040 - 2,270	(6)
2,270 - 2,470	(5)
2,470 - 2,860	(6)

Numbers in parentheses are numbers of occurrences.

Figure 1. Station locations in the MMS-supported investigation referred to as the North Gulf of Mexico Continental Slope Study (NGOMCS), conducted by LGL Environmental, Inc. Log transformations of polychaetes densities ($/m^2$) in the northern Gulf of Mexico (Hubbard 1995).

Considerable work has been initiated in the 1990's but little of it has been published yet. The *Instituto Ciencias del Mar y Limnologia* (ICMyL) of the *Universidad Nationale Autonoma de Mexico* (UNAM) initiated extensive studies in the southern Gulf of Mexico with the acquisition of the deep ocean research vessel *Justo Sierra*. Mexican biologists are conducting studies of megafauna, demersal fishes, macrofauna and meiofauna from the continental shelf down across the slope onto the Sigsbee Abyssal Plain. Stable carbon and nitrogen isotopes have been used to infer pathways through a benthic food chain (Soto and Escobar 1995). Studies in deep water across the *Cordilleras Mexicanas* or "Mexican Ridges" of the upper continental rise out onto the

Sigsbee Abyssal Plain, they have identified regions that contain enhanced biomass under surface water masses characterized by accelerated rates of primary production (Escobar-Briones et al. 1997; 1999). Polychaetes dominated the infauna; they encountered a mid-slope maximum in abundance similar to that described in the northern Gulf (Pequegnat et al. 1990). The fauna could be partitioned into three groups that conformed to depth intervals of > 3 km, intermediate depths of 1.5 and 3 km, and shallow waters < 1.5 km. This contrasts with the view of Pequegnat et al. that the continental margin is characterized by 5 zones.

In 1997, a two-ship operation by Texas A&M and UNAM was conducted at a common station (25°15' N. Lat. x 93°26' W. Long.) on the northern Sigsbee Abyssal Plain, at a depth of 3.65 km. The *Justo Sierra* traveled up from Tuxpan, sampling along an east - west line across the *Cordillera Tampiquena* or Mexican Ridges (Escobar Briones et al. 1999), while the *Gyre* went due south out of Galveston, directly to the site. As this site is well within the Mexican Exclusive Economic Zone (EEZ), all sampling conducted from the *Gyre* had to be approved *a priori* by a suite of Mexican federal agencies. It is nestled within several nearby locations that Pequegnat sampled with the skimmer and bottom photography.

Total sediment community respiration was measured using a benthic lander containing a pair of automatically operated benthic incubation chambers used to measure fluxes of metabolic gases across the sediment water interface. The lander and its operation have been described previously by Rowe et al. (1997). Oxygen consumption by the bottom and its contained biota is calculated from the decline of oxygen within the chamber over time, the volume of the chamber and the area of the sea floor it covers. It was deployed once in the western Sigsbee Deep at a depth of 3.6 km.

Microbiota

The microbiota of the deep Gulf sediments is not well characterized. While direct counts have been coupled with some in situ and re-pressurized metabolic studies in other deep ocean sediments (Deming and Baross 1993), none has been made in the deep Gulf. Direct counts using a fluorescing nuclear stain have been made at several depths down the slope, thus allowing the bacterial biomass to be estimated from their densities and sizes (Cruz-Kaegi 1998). This indicated that the bacteria are the most important component of the functional biota in terms of biomass. Comparisons are not yet possible with other ocean basins. Cruz-Kaegi (1998) developed a budget for carbon cycling based on her estimates of biomass and metabolic rates in the literature. Her budget illustrates that on the deep slope of the Gulf, a large fraction of the organic carbon supplying the benthos with energy is cycling through the bacteria.

The Meiofauna

The extensive sampling by the Northern Gulf of Mexico Continental Slop (NGOMCS) discovered that the meiofauna appears to have a biomass that is higher than that of the macrofauna. The regressions of log10 numbers per m^2 and biomass (as micrograms wet weight per m^2) were the following:

Log density = 5.87 - 0.00018 (depth in meters), and

Log biomass = 6.4 - 0.0002 (depth in meters).

Unfortunately, no estimates of biomass were made directly. Biomass was estimated from conversion factors for each group in both the macrofauna and the meiofauna. The densities were 2.5 log units higher than macrofauna, which is not unusual. The biomass was 0.5 log units above macrofauna, which is a reversal with the relationship found in shallow water environments. If true, this confirms earlier studies in the Atlantic (Rowe et al. 1991) that implied that meiofaunal-sized organisms increase in importance (in terms of biomass) at deep-sea depths relative to the macrofauna. Cruz-Kaegi (1998) also observed this in her studies. No information is available on the species composition or diversity of the meiofauna of the deep Gulf.

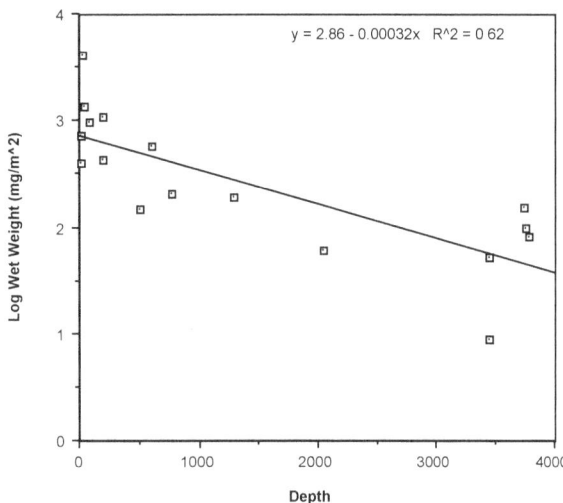

Figure 2. Density of macrofauna in the Gulf of Mexico (Rowe et al. 1974).

The Macrofauna

Abundance and Biomass: Anchor dredge samples across the Sigsbee Deep and van Veen grabs from the northern continental slope suggested that deep biomass in the Gulf was depauperate in numbers (Figure 1) and biomass (Figure 2), similar to that in other ocean basins, but that the mean size of the macrofauna was in general smaller than that in the Atlantic at similar depths (Rowe 1971; Rowe and Menzel 1971; Rowe et al. 1974; Rowe 1983). The log-normal relationship between biomass and depth has been confirmed now for numerous ocean basins (Rowe 1983), but the slope of the line (biomass as a function of depth) for the Gulf appeared to be steeper than that in most basins. It was suggested that this is due to the Gulf's low primary productivity.

Abundance values are a function of sieve size. The NGOMCS study on the slope used 300 micron sieves and the studies aboard *Justo Sierra* used 125 micron sieves. It might be expected that they both sampled far more individuals per square meter than the earlier studies by Rowe, who used a 420 micron sieve. The values for the NGOMCS study samples however ranged from 518 to 5369 ind/m^2, which is similar to that encountered by Escobar et al. in the southwestern Gulf and by Rowe et al. (1974) in the northern Gulf.

Regressions of biomass and abundance as a function of depth follow:

$$Log(10) \text{ ind/m}^2 = 3.52 - 0.000109 \text{ (depth in meters), and}$$

$$Log(10) \text{ microgram wet weight/m}^2 = 5.88 - 0.000227 \text{ (depth)}.$$

The polychaete densities have been plotted (Figure 3) as an example of information available from the NGOMCS study by LGL. Note that there is a gradual decline in abundances down the continental slope.

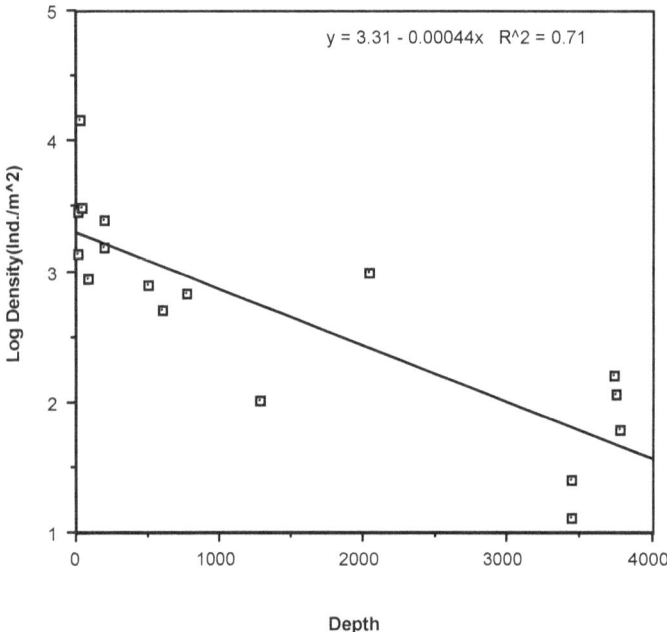

Figure 3. Biomass of macrofauna in the Gulf of Mexico (Rowe et al. 1974).

It has become clear that deep offshore abyssal plains may have somewhat fewer organisms than the Sigsbee Deep, but they are at much greater depth (Glover et al. in press). The deep Gulf has on the order of 150 to 500 ind/m², depending on sieve size, which taxa are included, and region studied. Escobar-Briones et al. (1999) for example suggested that the southern Gulf, in the region of the Bay of Campeche, had fewer macrofauna than the northern Gulf. They suggested that this was due to higher primary production that resulted from the interaction of warm eddies and shelf water.

The calculation of the diversity of the macrofauna was studied extensively in the NGOMCS study slope samples. The general pattern was a decline in diversity from the upper slope down to the lower slope, as reviewed by Lohse (1999). She suggested, by comparison with other studies on the outer shelf, that a "diversity maximum" was located on the upper slope. This contradicts earlier work in other basins which suggested that a diversity maximum is routinely encountered at 2 to 3 km depth. Hubbard (1995) also utilized the polychaete annelid fraction of the NGOMCS study to assess Gulf diversity. The polychaete fraction was used because it was the most abundant taxon (65%) and he was confident that he could separate them at the species level.

The studies of diversity used several measures. The diversity index referred to as the Information Function, or H'(s), was calculated from the polychaete data because it is simple to use, relatively independent of sample size, and has been used in numerous other studies, including Gulf of Mexico benthos. Fragments were not included in the analysis. The equation used follows:

$$H'(s) = -\Sigma (\ln p_i \times p_i),$$

where p is the proportion of each species, i, to the total population sampled, i through n. Also, Hurlbert's revision of the Sanders "rarefaction" curve, or Expected Species (per 100 individuals), was calculated and plotted by Lohse, directly from the NGOMCS study data reports, and by Hubbard from his own polychaete data.

In general the Gulf slope macrofauna is very speciose. Dominants are rare. Rare species are common. While shallow shelf studies in the Gulf typically encounter numerous individuals represented by 50 to 60 or so species, the slope data suggest that 100 or so species would be expected for a similar number of individuals. It has not yet been possible to compare the Gulf with other similar studies because of differences in technique.

There has been little attempt to uncover any seasonal variation in deep Gulf standing stocks or processes. Sets of samples along the central transect of the NGOMCS study were taken in spring and fall which suggested that some change in biomass could be observed in the animal abundances on the upper slope < 1 km depth (Figure 4). Polychaete abundance is twice as high in spring than in two fall periods. Species composition did not appear to change, just densities. The best description of this is in Hubbard (1995).

The Megafauna

Megafauna in the Gulf appears to occur in zones that are restricted to fairly predictable depth intervals (Pequegnat 1983; Pequegnat et al. 1990). These were designated 1) the Shelf/Slope Transition, down to about 500 m; 2) the Archibenthal Zone, Horizon A (500 to 775 m); 3) the Archibenthal Zone, Horizon B (800 to 1000 m); followed by the Upper Abyssal, Meso-Abyssal and Lower Abyssal Zones. These span the following depths: 1 to 2.3, 2.3 to 3.2, and 3.2 down to 3.8 km. These names were given earlier by Menzies et al. (1973) for similar zones encountered along several continental margins of the world oceans. In Pequegnat et al. (1990), percent similarities were presented as evidence that groups of fishes and megafauna in the region of the De Soto Canyon in the eastern Gulf occur in recurrent groups that could be considered zones. The rate of change in species composition of the bottom fishes also suggested that there are depth intervals that can be considered zones and other intervals that can be regarded as boundaries between zones. The authors mention the species that dominate each zone. Most of these can be found archived in the deep-sea systematic working collections at Texas A&M University. The predictability of zones such as this, regardless of what they are called, may be useful for understanding the effects of environmental variation, whether it is natural or the result of some alteration by human activities.

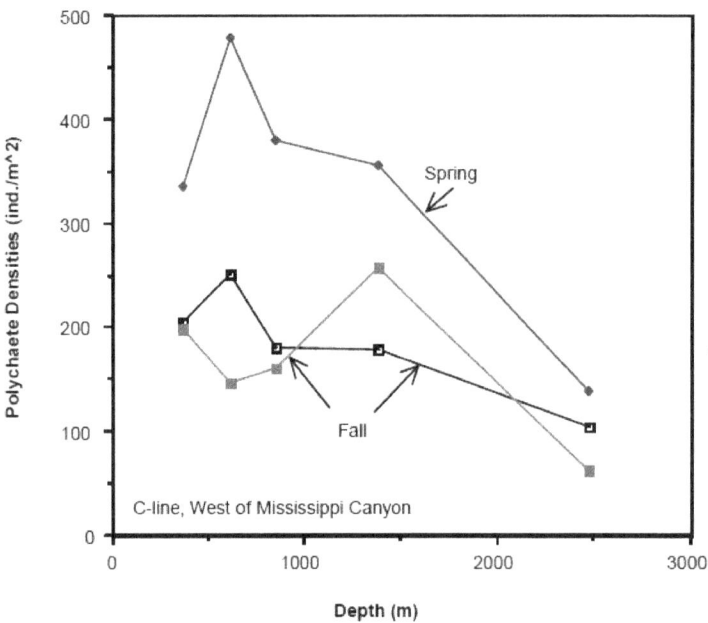

Figure 4. Comparison of polychaete densities at two different time periods on the upper slope of the northern Gulf of Mexico. (Data from Hubbard 1995).

The deep Gulf summary by Pequegnat (1983) suggested that megafauna on the abyssal plain was substantially reduced in both numbers and species compared to the continental slope. The megafauna was dominated by the carnivorous sea star *Dytaster insignis* and the surficial deposit feeding sea cucumber *Benthodytes typica*. Both of these species had wide bathymetric distributions that extended well up onto the continental slope. Other, less abundant megafauna species were also observed with some regularity. This included the brittle star *Ophiomusium planum*, which reached high densities in isolated locations when encountered. Others observed were the sea cucumber *Psychropotes semperiana* and the penaeid crustacean *Benthesicymus cereus/iridescens*. A number of other large crustacean species were observed in these earlier studies, but it is not clear if they were associated with the bottom or were captured up in the water column (*Nematocarcinus ensifer*, for example). There was little evidence that many demersal fishes extended out onto the abyssal plain. Pequegnat (1983) suggested that the "terminal predator" on the abyssal plain was the sea star *D. insignis*; he inferred this because the predatory sea stars increased in abundance at depth intervals over which the demersal fishes were declining.

The abundance of the dominant megafauna was estimated with two methods in the studies that Pequegnat (1983) has reviewed: multishot bottom photographic transects and use of the skimmer with an odometer wheel. The sea star *D. insignis*, according to Pequegnat, had mean densities of approximately 5 per hectare at depths of 3.6 km. *B. typica* reached similar values: from 4 to 7 individuals per hectare. Similar numbers were observed for the large crustaceans *B. cereus/iridescens*, *Nematocarcinus ensifer*, and the sea cucumber *P. semperiana*. It is not clear however if the crustaceans were caught on the bottom or in the water column; therefore they have been left out of the assessment of total benthic standing stocks.

Biomass was not measured in these earlier studies. However, many of the specimens are archived in a "systematics working collection" originally established by Willis and Linda Pequegnat on the Texas A&M campus. This enabled our review to find a representative number of the dominants, *B. typica* and *D. insignis*, and determine their mean biomass on preserved specimens approximately 30 years after they were captured. The *D. insignis* were dried specimens. They had a mean dry weight of 2.88 g per individual (σ=2.2, n=11). Mean disk diameter was 23.3 mm (σ=9, n=11). The holothurian *B. typica* individuals have been preserved in 70% ethanol and they were measured wet: mean=4.04 grams per individual (σ=1.4, n=13). They had a mean length of 6.6 cm (σ=1.1, n=13) and a diameter of 1.7 cm (σ=0.35, n=13). Thus, the sea star had a mean dry weight of 14.4 g per hectare and the holothurian had a mean wet weight of 22.2 g per hectare. The latter value would be equivalent to approx. 3.3 g per hectare dry weight (Rowe 1983). The two together would be equivalent to approximately 1.77×10^{-3} g dry weight m^{-2}.

Some megafauna are distributed in peculiar patches. The most consistent such animal patches in the Gulf appear to be the contagious distribution of the holothuroid *Peniagone* sp. These were rarely taken in trawls, but occurred as clumps of dozens of individuals all piled up, in the photographic surveys made in NGOMCS. The highest densities at a lowering were 1.6×10^{5} per hectare at a depth of 1.25 km. The reasons for the patchiness is not known. Other patches have been attributed to the proximity to fossil hydrocarbon seeps. Odd "reefs" of sponges are seen occasionally, but the causes of these accumulations are not known. Densities estimated with photography were always far higher than that estimated with the trawls.

Motile Scavengers

Little is known about deep-living scavengers in the GOM. On several occasions baited traps have been used to capture organisms. None of the information in these studies has been published however. In the 1997 A&M/UNAM studies with *Gyre* and *Justo Sierra*, a deep baited trap was deployed and a single species was captured: the cosmopolitan amphipod crustacean *Eurythenes grillus*. The numbers taken however appeared to be small (21 per 24 hour deployment) compared to similar trap deployments in other ocean basins. The significance of this group needs further study in the deep Gulf.

Community Metabolism

Sediment oxygen consumption has been measured at a single site on the deep abyssal plain in the eastern Gulf, and the rate measured was equivalent to approximately 6.7 mg C m^{-2}d^{-1} (Hinga et al. 1979). On the Sigsbee Abyssal Plain, an unpublished value of 4.1 mg C m^{-2}d^{-1} was measured. At this time it is impossible to know if this difference is real or simply due to a difference in gear used. In any case, the two values are not remarkably different from values at similar depths and temperatures elsewhere in the world's oceans. Values listed by Smith and Hinga (1983) for this depth range from a low of 2.6 up to a high of 25 mg C $m^{-2}d^{-1}$, and so our rate falls at the low end of the range.

Community Function – The Cycling of Organic Carbon

The information on the stock sizes and respiration rates of the biota tabulated have been put together into a carbon budget for the slope and abyssal plain (Cruz-Kaegi 1998). A budget of this sort allows a comparison of how carbon is both stored and cycled within an ecosystem. Values for the boxes (see presentation) are standing stocks in units of mg C m^{-2}, integrated to a depth of 20 cm, whereas the arrows are fluxes of carbon between the boxes, with units of mg C $m^{-2}d^{-1}$. The arrows that start as "little clouds" are inputs of organic carbon to the system whereas the arrows that end in clouds are losses of organic carbon from the system. Most of the losses are the remineralization of organic matter to CO_2. The input term, in this case, is estimated from the sum of the respiration loss terms by the community, plus long-term burial. An independent estimate of input can also be derived from measures of primary productivity in the overlying water and then applying the "Suess relationship" (Suess 1980) to estimate arrival of particulate matter at the sea floor. We assume the input varied by a factor of two or so over an annual cycle, with the highest productivity occurring during the winter when the phytoplankton biomass is highest (Muller-Karger et al. 1991).

Most of the loss terms (arrows terminated by cloud-like figures) are respiration. Respiration remineralizes organic carbon to metabolic CO_2. It is the largest consumer of organics in any typical benthic community and it can be estimated from oxygen fluxes. The sum of the aerobic respiration by the sediment dwelling organisms (bacteria, meiofauna and macrofauna) is equal to the value measured with the incubation chambers. This can be converted to carbon by assuming an RQ of 0.85 (moles of CO_2 produced per moles of O_2 consumed).

The respiration of the individual groups were estimated by Cruz-Kaegi independently from known size and temperature relationships in the literature (Mahaut et al. 1995). The regression produced by Mahaut et al. (1995) relating size and respiration rates was used to calculate a first-order respiration coefficient for the mean size of each size group.

The estimates of respiration for each size category allows us to further partition the flow of organic matter through the food web. The sum of the arrows entering each standing stock must equal the sum of the arrows leaving each stock, for the system to be in steady state. Given the information generated on the respiration rates of each of the components above, we can then calculate exchanges between the boxes to maintain steady state. This is a step-wise analysis that has been utilized previously on benthos over a broad range of latitudes (Cruz-Kaegi 1998), in addition to the Demerara Abyssal Plain (Rowe and Deming 1985) and the continental margin off NE Greenland (Rowe et al. 1997). The resulting solutions for the predator-prey relationships are not always unique. As indicated, they are based on reasonable inferences of how the size classes are most probably partitioning their resources.

Cruz-Kaegi's relationships are close to several of the measurements described in Relexans et al. (1996). A value for bacterial efficiency that is high tends to "conserve" carbon within the system, rather than burning it off as metabolic CO_2. The production of the bacteria must be held in check by bactivory to maintain steady state and this has been directed primarily into the meiofauna, with a small fraction into the macrofauna. It was assumed the meiofauna had a 10% growth efficiency and thus the loss terms are respiration and predation by macrofauna, as indicated. The total meiofauna demand must therefore be met by bactivory and direct sediment consumption. Macrofauna fluxes are calculated in a similar fashion.

The megafauna was assumed to be growing at a very slow rate and this is transferred by predation to the fishes. Fish growth is assumed to be zero; thus, the megafauna growth rate is equal to fish respiration.

The standing stock of the organic matter was calculated from the concentration of organic carbon (0.6%), the porosity (85%) and the density of the sediment. Thus, organic carbon in the surface 5 cm of sediment is 3.6 g C m^{-2}-5 cm depth. The burial rate was estimated from longterm burial rates (William R. Bryant, pers. comm.), as follows. Approximately 1 to 2 meters of Holocene pelagic sediments, composed primarily of foram ooze, are spread rather uniformly over the entire Sigsbee Deep. On the continental margins the thickness of the Holocene material is 3 to 4 meters deep. Thus, the rate of accumulation on geologic time scales are ca. 100 to 200 cm per 13,000 years; at the time scales of our budget this is ca. 0.015 cm y^{-1}. Multiplying this by the concentration in a 1 cm thick layer (225 mg C m^{-2}-cm depth) gives a burial rate of ca. 3.5 mg org. carbon m^{-2} per year, or 0.0095 mg C $m^{-2}d^{-1}$, the appropriate units in the budget. This value is comparable to the low flux rates estimated for the larger sized groups of organisms (megafauna and fishes).

Comparison of the Gulf Benthos with Other Oceans

The standing stocks of the principal components of the benthic community have been measured together for the first time at a site in the Sigsbee Deep in the northwest Gulf of Mexico. The relationship of bacteria numbers in abyssal sediments has been regressed on depth, organic carbon concentrations and POC fluxes to the sea floor by Deming and Baross (1993). They found that the best predictor of log_{10} total bacterial biomass was $log_{10}POC$ fluxes. If we assume that our "input" term in our budget, estimated from the sum of the measured or estimated carbon demand, is equal to the POC fluxes measured in sediment traps in their study, then we can use this term in their regression. Our value for biomass (408 mg C m^{-2}-20 cm depth) suggests, based on their regression, that the POC flux should be on the order of 24 mg C $m^{-2}d^{-1}$.

The abundance and biomass of the macrofauna was somewhat higher than previously estimated in the southern Gulf of Mexico (Rowe and Menzel 1971). Escobar-Briones et al. (1999) made comparisons with several similar ocean basins and noted that abundances were higher in the northern Gulf than in the Bay of Campeche. We suggest that this area experiences intensified surface primary production that results from the interaction of warm eddies (Muller-Karger et al. 1991). The differences with the measurements in the southern Gulf presented by Rowe and Menzel (1971) may be due to gear: the earlier studies used a semi-quantitative anchor dredge and not a spade or box corer.

The biomass of the macrofauna was not statistically separable from a general mean for the "global" ocean at a depth of 3.7 km, based on more than 700 values measured with a wide variety of sampling gear (Rowe 1983). A regression of $\log_{10}$ biomass as wet preserved weight as a function of depth in meters suggests that biomass at 3.7 km depth at the deep end of the Gulf regression line would be expected to be ca. 0.64 g m^{-2}.

A fair number of trawls have been taken across extensive areas of the continental margin in both the NW Atlantic (Haedrich and Rowe 1977; Haedrich et al. 1980), and the NE Atlantic (Haedrich and Krefft 1978; Lampitt et al. 1986), thus making it possible to compare densities and biomass of megafauna and fishes with the Gulf. Lampitt et al. (1986) plotted $\log_{10}$ biomass of total invertebrate megafauna as a function of depth in the NE Atlantic near the Porcupine Bight. Their regression line predicts that 0.31 g wet weight m^{-2} should be encountered at 3.7 km depth. Haedrich et al. (1980) measured a wet preserved weight of ca. 0.08 g m^{-2} of fishes and 0.05 g m^{-2} of megafaunal invertebrates (echinoderms and crustaceans) between depths of 3.2 and 3.7 km, suggesting that the abundance and biomass of these groups is lower than in the Porcupine Bight of the NE Atlantic. The lower value above is higher than we have estimated for similar depths in the Gulf (0.006 g wet preserved weight m^{-2}= ca. 0.155 mg C m^{-2}); thus, compared to the Atlantic, the deep Gulf megafauna appears to be relatively depauperate. The extensive data from both the east and west sides of the North Atlantic had ranges of as much as three orders of magnitude at any given depth.

The diversity of the polychaete fraction confirms that the Gulf of Mexico harbors a diverse fauna, as is typical for much of the deep-sea floor (Rex 1983). The numbers of species, H'(s) and the expected number of species per hundred individuals however are rather low compared to values on the upper continental slope and outer continental shelf (Lohse 1999). This pattern relative to depth is different from that in other basins: maximum diversity, regardless of "index" used, is found on the lower slope or the upper rise. The values we encountered were somewhat lower than that in other basins. Thus, a preliminary conclusion is that the pattern in the Gulf is different from most basins, with maximum diversities at the shallower depths of the upper slope and outer shelf (100 to 1000 m), rather than at the slope base (2 km to 3 km), as described by Rex (1983). The apparent decline with depth might be due to the limiting sill depth, the variability of the deep water environment (seeps from fossil hydrocarbons buried deep in the sediments, temperature variations due to warm eddies originating in the loop current in the eastern Gulf, nepheloid layers associated originally with the Mississippi River, sediment slumps and turbidity flows), or inadequate sample size. The latter artifact cannot be discounted because we continued to add equal numbers of previously un-encountered species with each additional box core, thus suggesting that many more species were present than we had actually sampled. This question can only be answered by more intense sampling.

The abyssal plain of the Gulf differs from those in the major ocean basins because it is shallower (3.6 to 3.8 km) and warmer (4.2°C). Thus, one might expect that the characteristics of communities of organisms might differ from "typical" oceanic abyssal plains.

Summary

The study of the deep benthos of the Gulf of Mexico can be divided into three eras. Early studies prior to the 19th century were exploratory zoogeographic investigations. In the 1960's Willis Pequegnat initiated modern investigations of the standing stocks of the megafauna. In the 1980's the MMS supported an investigation of the continental slope that encompassed state-of-the-art quantitative sampling of the meiofauna, macrofauna and megafauna. Metazoan biomass is dominated by the meiofauna. The eastern Gulf is not similar in species composition to the western Gulf and this can be attributed to a difference in substrate. The western sediments are terrigenous whereas the eastern Gulf substrate is carbonate. Faunal diversity of the macrofauna is high but it is not characterized by a mid-depth maximum. The low biomass of the macrofauna and the megafauna suggest that the biota is food limited. Budgets of carbon cycling on the slope suggest that most of the organic matter is cycled through the smaller animals of the communities.

References

Agassiz, A. 1888. Three Cruises of the Blake. Mus. Comp. Zool., Harvard.

Boland, G. and G. Rowe. 1991. Deep-sea benthic sampling with the GOMEX box corer. Limnol. and Oceanogr. 36(5):1015-1020.

Booker, R. 1971. Some aspects of the biology and ecology of the deep-sea echinoid *Phormosoma placenta* Wyv. Thompson. M.S. thesis, Texas A&M University.

Bright, T. 1968. A survey of the deep-sea bottom fishes of the Gulf of Mexico below 350 meter. Ph.D. dissertation, Dept. of Oceanography, Texas A&M University.

Carney, R. 1971. Some aspects of the ecology of *Mesothuria lactea* Theel, a common bathyal holothurian in the Gulf of Mexico. M.S. thesis, Texas A&M University.

Cruz-Kaegi, M. E. 1998. Latitudinal variations in biomass and metabolism of benthic infaunal communities. Ph.D. dissertation, Texas A&M University.

Deming, J. and J. Baross. 1993. The early diagenesis of organic matter: Bacterial activity. Pp. 119-144, in: M. Engel and S. Macko (Eds.), Organic Geochemistry. Plenum, New York.

Escobar-Briones, E. G., M. Signoret and D. Hernandez. 1999. Variacion de la densidad de la infauna macrobentica en un gradiente batimetrico: oeste del Gulfo de Mexico. Ciencias Marinas 25:1-20.

Escobar-Briones, E. and L. Soto. 1997. Continental shelf benthic biomass in the western Gulf of Mexico. Cont. Shelf Res. 17:685-604.

Firth, R. 1971. A study of the deep-sea lobsters of the families Polychelidae and Nephropidae (Crustacea, Decapoda). Ph.D. dissertation, Texas A&M University.

Gallaway, B. (Ed.). 1988. Northern Gulf of Mexico Continental Slope Study, Final Rept.: Year 4. Vol. II: Synthesis Rept. Final report submitted to the Minerals Management Service, New Orleans, LA Contract No. 14-12-0001-30212..

Gallaway, B., L. Martin and R. Howard. (Eds.). 1988. Northern Gulf of Mexico Continental Slope Study. Annual Rept.: Year 3. Vol. I-III: Executive Summary, Technical Report and Appendices. Submitted to the Minerals Management Services, New Orleans, LA. Contr. No. 14-12-0001-30212. OCS Study/MMS 87-0061.

Galtsoff, P. 1954. Gulf of Mexico-Its Origin, Waters and Marine Life. U.S. Dept. of Int., Fish and Wildlife Service, Fishery Bull. 89.

Geyer, R. 1970. Preface, in Contribution on the Biology of the Gulf of Mexico. W. Pequegnat and F. Chace, Jr. (Eds.).

Glover, A., G. Paterson, B. Bett, J. Gage, M. Sibuet, M. Sheader and L. Hawkins. In press. Patterns in polychaete abundance and diversity from the Madeira Abyssal Plain, northeast Atlantic. Deep-Sea Research.

Haedrich, R. L. and G. Krefft. 1978. Distribution of bottom fishes in the Denmark Strait and Irminger Sea. Deep-Sea Research 25:705-720.

Haedrich, R. L. and G. Rowe. 1977. Megafaunal biomass in the deep sea. Nature 269:141-142.

Haedrich, R. L., G. Rowe and P. T. Polloni. 1980. The megafauna in the deep-sea south of New England. Mar. Biol. 57:165-179.

Hinga, K., J. McN. Sieburth and G. R. Heath. 1979. The supply and use of organic material at the deep-sea floor. Journal of Marine Research 37:557-579.

Hubbard, F. 1995. Benthic polychaetes from the northern Gulf of Mexico continental slope. Ph.D. dissertation, Texas A&M University.

James, B. 1972. Systematics and biology of the deep-water Palaeotaxodonta (Mollusca: Bivalvia) from the Gulf of Mexico. Ph.D. dissertation, Texas A&M University.

Kennedy, E. A., Jr. 1976. A distribution study of deep-sea macrobenthos collected from the western Gulf of Mexico. Ph.D. dissertation, Texas A&M University.

Lampitt, R., D. Billett and A. Rice. 1986. Biomass of the invertebrate megabenthos from 500 to 4100 m in the northeast Atlantic Ocean. Marine Biology 93:69-81.

Lohse, A. 1999. Variation in species diversity within macrobenthic invertebrate communities in the western Gulf of Mexico. M.S. thesis, Texas A&M University.

Mahaut, M., M. Sibuet and Y. Shirayama. 1995. Weight-dependent respiration rates in deep-sea organisms. Deep-Sea Research 42:1575-1582.

Menzies, R. J., R. Y. George and G. Rose. 1973. Abyssal environment and ecology of the world oceans. John Wiley and Sons, Inc., New York, 448 p.

Muller-Karger, F., J. Walsh, R. Evans and M. Meyers. 1991. On the seasonal phytoplankton concentration and sea surface temperature cycles of the Gulf of Mexico. Journal of Geophysical Research 90:12645-12665.

Pequegnat, L. 1970. A study of deep-sea caridean shrimps (Crustacea: Decapoda: Natantia) of the Gulf of Mexico. Ph.D. dissertation, Texas A&M University.

Pequegnat, W. 1983. The ecological communities of the continental slope and adjacent regimes of the northern Gulf of Mexico. Report by TerEco Corp. to the Minerals Management Service on contract AA851-CTI-12. 398 p. + App.

Pequegnat, W., T. Bright and Bela James. 1970. The benthic skimmer, a new biological sampler for deep-sea studies. Pp. 17-20, in: W. Pequegnat and F. Chace, Jr. (Eds.), Contributions on the Biology of the Gulf of Mexico. Gulf Publishing, Houston.

Pequegnat, W., B. Gallaway and L. Pequegnat. 1990. Aspects of the ecology of the deep-water fauna of the Gulf of Mexico. American Zoologist 30:45-64.

Pequegnat, W., B. James, A. Bouma, W. Bryant and A. Fredericks. 1972. Photographic study of deep-sea environments of the Gulf of Mexico. Pp. 67-218, In: V. Henyard and R. Rezak (Eds.), Contributions on the geological oceanography of the Gulf of Mexico. Texas A&M University Oceanography Studies. Vol. 3, Gulf Publ. Co., Houston, TX.

Rex, M. 1983. Geographic patterns of species diversity in the deep-sea benthos. Pp. 453-472, In: G. Rowe (Ed.), Deep-Sea Biology, Vol. 8, The Sea. Wiley, Interscience.

Rayburn, R. 1975. Food of deep-sea fishes of the northwestern Gulf of Mexico. M.S. thesis, Texas A&M University.

Relexans, J.-C., J. Deming, A. Dinet, J.-F. Gaillard and M. Sibuet. 1996. Sedimentary organic matter and micro-meiobenthos with relation to trophic conditions in the tropical northeast Atlantic. Deep-Sea Research I 43:1343-1368.

Roberts, T. 1970. A preliminary study of the family Penaeidae and their distribution in the deep water of the Gulf of Mexico. M.S. thesis, Texas A&M University.

Roberts, T. 1977. An analysis of deep-sea benthic communities in the northeast Gulf of Mexico. Ph.D. dissertation, Texas A&M University.

Rowe, G. 1971. Benthic biomass and surface productivity. Pp. 441-454, In: J. Costlow (Ed.), Fertility of the Sea, Vol. II. Gordon and Breach, New York.

Rowe, G. 1983. Biomass and production in deep-sea macrobenthos. Pp. 97-121, in: G. Rowe (Ed.), Vol 8, The Sea, Deep-Sea Biology. Wiley, New York.

Rowe, G. and D. W. Menzel. 1971. Quantitative benthic samples from the deep Gulf of Mexico with some comments on the measurement of deep-sea biomass. Bull. Mar. Sci. 21(2):556-566.

Rowe, G. and J. Deming. 1985. The roles of bacteria in the turnover of organic carbon in deep-sea sediments. Journal of Marine Research 43:925-950.

Rowe, G., P. T. Polloni and S. B. Hornor. 1974. Benthic biomass estimates from the northwestern Atlantic Ocean and the northern Gulf of Mexico. Deep-Sea Res. 21:641-650.

Rowe, G., M. Sibuet, J. Deming, A. Khripounoff, J. Tietjen, S. Macko and R. Theroux. 1991. "Total" sediment biomass and preliminary estimates of organic carbon residence time in the deep-sea benthos. Marine Ecology Progress Series 79:99-114.

Rowe, G., G. Boland, E. G. Escobar Briones, M. E. Cruz-Kaegi, A. Newton, D. Piepenburg, I. Walsh and J. Deming. 1997. Sediment community biomass and respiration in the northeast water polynya, Greenland: a numerical simulation of benthic lander and spade core data. Journal of Marine Systems 10:497-515.

Smith, K. L., Jr. and K. Hinga. 1983. Sediment community respiration in the deep sea. Pp. 331-370, in: G. Rowe (Ed.), Deep-Sea Biology, The Sea, Vol. VIII. Wiley, New York.

Smith, K. L., Jr. 1992. Benthic boundary layer communities and carbon cycling at abyssal depths in the central North Atlantic. Limnology & Oceanography 37:1034-1056.

Soto, L. and E. Escobar. 1995. Coupling mechanisms related to benthic production in the southwestern Gulf of Mexico. Pp. 233-242, in: A. Eleftheriou, A. Ansell and C. Smith (Eds.), Proc. 28th Eur. Mar. Biol. Symp. Olsen and Olsen, Copenhagen.

Suess, E. 1980. Particulate organic carbon flux in the oceans-surface productivity and oxygen utilization. Nature 288:260-263.

The Benthos of the Deep Gulf of Mexico
Click on the title to view the presentation.

The Benthos of the
Deep Gulf of Mexico

Gilbert T. Rowe
Texas A&M University

Overview of Fisheries of the Deep Gulf of Mexico

Randy E. Edwards* and Kenneth J. Sulak
U.S. Geological Survey, Biological Resources Division, Florida Caribbean Science Center, Gainesville, Florida
*University of South Florida, College of Marine Science, St. Petersburg, Florida

Introduction

This paper summarizes information on important Gulf of Mexico (GOM) fish and fisheries in deepwater (depth >1,000 ft or 305 m) areas into which petroleum exploration and production has recently expanded (Baud et al. 2000). It also discusses the potential for GOM deepwater petroleum structures (DPSs) to affect fisheries by acting as fish aggregating devices (FADs) and how DPSs' effects on fish and fisheries are very different from those of petroleum structures in shallower waters.

Species

The most important fish in the deepwater environment are open-ocean, pelagic species, which the National Marine Fisheries Service (NMFS) terms highly migratory species (HMS). The most important HMS are marlins, sailfish (*Istiophorus platypterus*), swordfish (*Xiphias gladius*), tunas, and certain pelagic shark species. Dolphin (*Coryphaena hippurus*), wahoo (*Acanthocybium solandri*) are also important pelagic species in GOM deepwaters, although they do not fall within the NMFS HMS category. A few coastal pelagic species such as amberjacks (*Seriola* sp.), cobia (*Rachycentron canadum*), and king mackerel (*Scomberomorus cavalla*) are sometimes present in deepwater environments but usually are not as abundant as in shallower, shelf environments. Most benthic fish species, like red snapper (*Lutjanus campechanus*), red grouper (*Epinephelus morio*), gag (*Mycteroperca microlepis*), etc., that are important in shallower waters of the continental shelf are not found in deepwater environments. A few deepwater groupers, including misty grouper (*Epinephelus mystacinus*), yellowedge grouper (*Epinephelus flavolimbatus*), and snowy grouper (*Epinephelus niveatus*) extend into deepwaters of the GOM.

The GOM deepwater environments serve as habitat for one important invertebrate species, the royal red shrimp (*Hymenopenaeus robustus*). Essential habitat for royal red shrimp has been designated by NMFS[1] as the upper regions of the continental slope from 590 ft (180 m) to about 2,395 ft (730 m) with specific types of sediments. Trawling for royal red shrimp in the northern GOM extends to depths of around 1,000 ft (305 m) and can be expected to go deeper in the future (D. B. Snyder, personal communication).

---

[1] Source: NMFS Essential Fish Habitat Designations for South Atlantic Fishery Management Plans
http://www.nmfs.noaa.gov/habitat/habitatprotection/images/SAFMC.pdf

Commercial Fisheries

Commercial finfish fisheries in the deepwater GOM consist almost exclusively of longline fisheries that target yellowfin tuna (*Thunnus albacares*) or swordfish, using drifting lines that typically are up to 40 mi (~65 km) long with up to about 1,000 baited hooks (Berkeley and Edwards 1997). Over the period of 1981-2000, the pelagic longline fishery had an average annual value of $16.6 million, of which yellowfin tuna accounted for $11.6 million and swordfish for $3.4 million. During that period, yellowfin tuna accounted for 71% of the catch, swordfish 20%, bluefin tuna (*Thunnus thynnus*) 5%, dolphin 4%, and blackfin tuna (*Thunnus atlanticus*) <1%. A host of miscellaneous species is also caught incidentally by longlines.

A Japanese high-seas longline fishery for yellowfin tuna operated in the GOM from 1957 until 1981 (Wilson 1988). A domestic GOM longline fleet was established in the early 1980's, and yellowfin tuna catches peaked at 9.2 million lb with dockside value of $19.2 million in 1992[2]. Landings declined substantially over the next few years, and recently (1996-2000), GOM yellowfin tuna landings have averaged 4.7 million lb and $11.1 million. This is a small fraction of the total Atlantic yellowfin tuna landings (most taken in the eastern Atlantic) of around 300 to 350 million lb/yr during the same period. Bluefin tuna (*Thunnus thynnus*) catches (incidental or sometimes targeted) were once an economically important component of the fishery, due to the very high prices bluefin tuna commanded in the Japanese market for sashimi. Bluefin tuna landings peaked in 1988 at 0.3 million lb and $4.1 million ($13.54/lb), but declined greatly thereafter in response to catch restrictions and declining prices (recently $4-5/lb) brought about as the market found new sources for sashimi-grade tuna in other parts of the world.

Analyses of the extensive NMFS database of longline fishing locations and catches (Brown and Scott 2001; Snyder et al. 2001) show that considerable longline effort and catch of a variety of species occurs within areas of the continental slope into which deepwater petroleum activities have expanded or are expected to expand. Beyond the longline catch data, little is known about the distributions, movements, and migrations of pelagic fishes in the GOM. There have been no extensive tagging/recapture studies of any species. However, limited data from yellowfin tuna tagging and recapture (Ortiz 2001) suggests, from the fact that yellowfin tagged in the GOM and from other parts of the western Atlantic were recaptured in the large commercial fishery in the eastern Atlantic, that GOM yellowfin tuna are part of single Atlantic stock. The extent to which seasonal or annual variation in yellowfin migration patterns affect abundance in the GOM is unknown. It is known, however, that the GOM serves as the only known spawning grounds for the western Atlantic stock of giant bluefin tuna (Nemerson et al. 2000). Bluefin tuna spawn in the GOM during late spring and then migrate out of the GOM through the Straits of Florida in May or June to summer feeding grounds off the U.S. northeastern coast and the Canadian Maritimes. The western Atlantic stock of bluefin has declined to such low levels that some observers have suggested that it is threatened with extinction (Safina 1993). Skipjack tuna (*Katsuwonus pelamis*) is one of the most abundant tuna species in most oceans, but information about its abundance and distribution in the GOM is limited to observations that commercially exploitable quantities may exist there (Sakagawa 1986) and the fact that they have been caught in commercial quantities in pole-and-line fisheries around fish aggregation devices (FADs)

---

[2] Commercial Landings Source: NOAA Fisheries http://www.st.nmfs.gov/st1/commercial/

deployed off the northern coast of Cuba (Martin 1999). However, U.S. GOM catch of skipjack tuna has been extremely small.

Recreational Fisheries

Recreational fisheries are also important in deepwater areas of the GOM. However, it is not possible to partition deepwater fisheries from the overall offshore recreational fisheries that include areas shallower than 1,000 ft. The overall GOM offshore recreational landings (1991-2000 average[2]) are dominated by dolphin (85%, 4.4 million lb), blackfin tuna (13%, 0.6 million lb) and yellowfin tuna (5%, 0.2 million lb). However, most offshore anglers target billfish, often for catch and release. Unfortunately, these fisheries have not been precisely quantified with regard to effort, catch or value. The NMFS recreational billfish survey (RBS) documented about 16,000 to 27,000 trolling hours per year for the Gulf of Mexico from 1988 through 1995 (NMFS 1997). The RBS GOM billfish catch per unit effort (CPUE) for 1994 was 3.0 billfish per 100 hr, which is comparable to CPUE in the Bahamas (3.2) and close to that for the Caribbean (3.8) and the Atlantic north of Florida (4.2). The 1994 GOM catch was comprised of about 54% blue marlin (*Makaira nigricans*), 35% white marlin (*Tetrapturis albidus*), and 11% sailfish (NMFS 1997). Overall, a large, productive, and economically valuable recreational fishery takes place in GOM deepwater areas.

GOM DPSs as Fish Aggregating Devices

The issue of whether deepwater petroleum structures might act as fish aggregating devices was initially raised in a prior workshop on deepwater oil and gas development (Carney 1997). In response, the U.S. Geological Survey (USGS) researched available scientific literature and information on FADs, compiled a bibliography, and provided synthesis and analysis pertinent to GOM DPSs functioning as FADs (Edwards and Sulak 2002). Much of the following discussion is taken from that report.

The literature on FADs is relatively limited and recent, and therefore Edwards and Sulak (2002) concluded that it is inadequate for extrapolation and prediction of GOM DPS FADs effects. However, a number of conclusions were possible. One of the most important is that relationships between deepwater petroleum structures and fish/fisheries are very different from those for structures in shallower waters. In deepwater environments, completely different biotopes, species, trophic linkages, and fisheries are involved. Hence, the fishery resource management issues for DPSs are very different from those that have been studied so intensively for shallower rigs. Therefore, if the effects and impacts of GOM deepwater petroleum structures are to be understood, direct study of DPSs will be needed.

Various types of floating structures, both drifting and anchored, have been deployed in open-ocean environments to attract and aggregate fish for commercial and recreational purposes. FADs have been used around the world, and the primary species that are associated with FADs are tunas, especially yellowfin tuna, skipjack tuna and bigeye tuna (*Thunnus obesus*). FADs have been found to be so effective that most of the world commercial tuna catch is now taken from around FADs (Fonteneau et al. 2000). Additionally, a suite of other pelagic fishes has been documented to also associate with FADs. They include dolphin, wahoo, blackfin tuna, blue marlin, bluefin tuna, and certain shark species.

The reasons why tunas and other fishes aggregate around FADs are not well understood. Nine reasonable hypotheses for this phenomenon have been proposed (Freon and Misund 1999). Only one of these, that fish aggregate because a concentrated food supply exists around FADs, has been discounted (Freon and Daghorn 2000). This is particularly important in that it indicates that DPSs would not enhance fish production, as has been proposed for shallow rigs, and would result only in attraction. Recently, the remaining eight hypotheses have been consolidated into four hypotheses (Freon and Daghorn 2000). They are that fish aggregate around objects because: 1) objects provide shelter from predation, 2) objects provide a spatial reference in an otherwise featureless environment, 3) drifting objects are indicators of productive areas for feeding, and 4) objects provide an opportunity for individuals or small schools to meet and form larger schools. The factors that determine the degree to which an object aggregates fish remain unknown. After an intensive study, Hall et al. (1992) concluded that as long as an object had a minimum dimension of one meter, they could find no other factors that were statistically related to the abundance of the aggregated fish. Therefore, it is impossible to understand or predict the aggregating effect of diverse and structurally complex petroleum structures.

However, it is known from acoustic tracking studies of yellowfin and bigeye tuna, that tuna movements are greatly influenced by the presence of FADs and that the radius of influence of a structure is on the order of 5 nmi (9 km) (Holland et al. 1990; Holland 1996; Marsac and Cayre 1998). Furthermore, existence of a network of structures can affect movements. Presence of four or five FADs in a 27 nmi (50 km) x 27 nmi area reduces skipjack tuna movement out of an area by 50% (Kleiber and Hampton 1994). Such a network of structures already is in place in the deepwater GOM and is equivalent, in area and number of structures, to the FADs network that has been deployed around the Hawaiian Islands to provide recreational and artisanal fishing opportunities (Holland et al. 2000).

Review of existing literature and information on FADs (Edwards and Sulak 2002) indicates that GOM DPSs may have positive or negative effects. Possible positive effects include increased catchability of fish for recreational and commercial fishermen, increased efficiency, and bycatch reduction under some circumstances. Possible negative effects include shift to smaller fish and resultant impacts on fishery yields, disruption of normal movements, impacts on spawning, increased fishing mortality, increased bycatch, user group conflicts, and fishery management problems.

Unfortunately, existing FADs literature and science is inadequate for predicting which of the positive or negative effects will occur. Some of the potentially critical issues include possible effects of GOM DPSs on valuable yellowfin tuna fisheries (e.g., through development of commercial fishing for yellowfin tuna around DPSs), development of new fisheries (e.g., for skipjack tuna), and impacts on the already critically low stocks of bluefin tuna (Safina 1993) (e.g., by attracting them away from optimal spawning areas (Lamkin et al. 2001) and subjecting them to increased bycatch).

Numerous reports of GOM DPSs already having FADs effects have come to light recently. Recreational fishermen have reported that deepwater oil and gas structures attract pelagic fish like yellowfin tuna, blue marlin, wahoo, dolphin, blackfin tuna, and sharks (Sloan 2001). One magazine article described deepwater rigs as follows: "These offshore rigs act as fish-attracting devices, supporting whole ecosystems, pulling in great numbers of pelagics and smaller game fish" (Husser 1997). Another article (Olander 2001) described catches of deepwater grouper and other bottom fish near a GOM DPS.

Research Needs

Given that GOM DPSs are already acting as FADs and that existing scientific information is inadequate for understanding or predicting how increasing numbers of DPSs will affect GOM fish and fisheries, direct field studies are recommended (Edwards and Sulak 2002). A number of research approaches have been used to study FADs. They include: tagging, hydroacoustic survey, acoustic telemetry/tracking, acoustic telemetry using automated receivers, fishery dependent catch surveys, fishery independent hook-and-line fishing, experimental net fishing, and modeling. Of these, acoustic telemetry/tracking (e.g., Holland et al. 1990; Holland 1996; Marsac and Cayre 1998; Dagorn et al. 2000) and hydroacoustic survey (e.g., Josse et al. 1998; Josse et al. 2000) have been most effective in elucidating fish/FADs relationships. In the acoustic tracking studies, the fish were tagged with ultrasonic transmitters and were tracked for extended periods to determine their movements relative to FADs. Additionally, GOM DPSs offer the opportunity to deploy automated acoustic receiver systems (e.g., Klimley and Holloway 1999) that can detect the periodic return of sonic-tagged fish. In the hydroacoustic studies, sonar instruments have been used to observe horizontal and vertical distributions of fish relative to FADs. Most recently, sonic tagging has been combined with hydroacoustic survey to allow individual fish to be recognized and species and sizes to be determined (Josse et al. 1998). State-of-the-art multibeam, three-dimensional, computerized sonar systems (Gerlotto et al. 1999) could, in the future, provide detailed information about fish distributions around DPSs and other structures.

Acknowledgments

Craig Brown, NMFS, and David Snyder, Continental Shelf Associates, generously provided copies of GOM longline catch and effort maps. Gary Brewer, USGS Eastern Regional Office, greatly aided and supported this work.

References

Baud, R. D., R. H. Peterson, C. Doyle and G. E. Richardson. 2000. Deepwater Gulf of Mexico: America's emerging frontier. U.S. Department of Interior, Minerals Management Service, Gulf of Mexico OCS Regional Office, New Orleans, LA.

Berkeley, S. A. and R. E. Edwards. 1997. Factors affecting billfish capture and survival in longline fisheries: potential application for reducing bycatch mortality. Col. Vol. Sci. Pap. ICCAT, 48:255-262.

Brown, C. and G. Scott. 2001. Tuna in the Gulf of Mexico. Abstract of paper presented at "Rigs and FADs" special session of the American Fisheries Southern Division Meeting, January 22-25, 2001, Jacksonville, FL.

Carney, R. S. 1997. Workshop on environmental issues surrounding deepwater oil and gas development: Final report. OCS study MMS 98-0022. U.S. Dept. Interior, Minerals Management Service, Gulf of Mexico OCS Region, New Orleans, LA. 163 pp.

Dagorn, L., E. Josse and P. B. Bach. 2000. Modeling tuna behaviour near floating objects: from individuals to aggregations. Aquat. Living Resour. 13:203-211.

Edwards R. E and K. J. Sulak. 2002. Potential for Gulf of Mexico deepwater petroleum structures to function as fish aggregating devices (FADs) – Scientific information summary and bibliography. Final Project Report, U.S. Department of Interior, Geological Survey, USGS BSR 2002-0005 and Minerals Management Service, Gulf of Mexico OCS Region, New Orleans, Ala, OCS Study MMS 2002-39. 261 pp.

Fonteneau, A., P. Pallares, and R. Pianet. 2000. A worldwide review of purse seine fisheries on FADs. Pp. 15-35 in: J.-Y. Le Gall, P. Cayre and M. Taquet (Eds.), Pêche thonière et dispositifs de concentration de poissons. (Tuna fishing and fish aggregation devices.). Actes Colloque I FREMER, 28, Brest Cedex.

Freon, P., and O. A. Misund. 1999. Dynamics of pelagic fish distribution and behavior: effects on fisheries and stock assessment. Fishing News Books, London.

Freon, P. and L. Dagorn. 2000. Associative behaviour of pelagic fish: facts and hypotheses. Pp. 483-491 in: J.-Y. Le Gall, P. Cayre and M. Taquet (Eds.), Pêche thonière et dispositifs de concentration de poissons. (Tuna fishing and fish aggregation devices.). Actes Colloque I FREMER, 28, Brest Cedex.

Gerlotto, F., M. Soria, and P. Freon. 1999. From 2D to 3D: methodology for a new approach in fisheries acoustics. Can. J. Fish. Aquat. Sci. 56:6-12.

Hall, M., C. Lennert, and P. Arenas. 1992. The association of tunas with floating objects and dolphins in the eastern Pacific Ocean. III: Characteristics of floating objects and their attractiveness for tunas. In: M. Hall (Ed.), The association of tunas with floating objects and dolphins in the eastern Pacific Ocean. Inter-American Tropical Tuna Commission, La Jolla, CA.

Holland, K. N. 1996. Biological aspects of the association of tunas with FADs. SPC FAD Inf. Bull. 2:2-7.

Holland, K. N., R. W. Brill and R. K. C. Chang. 1990. Horizontal and vertical movements of yellowfin and bigeye tuna associated with fish aggregating devices. Fish. Bull. 88:493-507.

Holland, K. N., A. Jaffe and W. Cortez. 2000. The Fish Aggregating Device (FAD) system of Hawaii. Pp. 55-62 in: J. Y. Le Gall, P. Cayre and M. Taquet (Eds.), Pêche thonière et dispositifs de concentration de poisons. (Tuna fishing and fish aggregation devices.). Actes Colloque I FREMER, 28, Brest Cedex.

Husser, R. 1997. Twilight zone tuna. Sportfishing Magazine Sept./Oct. 1997 <http://www.sportfishingmag.com/SF_Main/1,3214,5-1-1-21624-335-129,00.html>.

Josse, E., P. Bach, and L. Dagorn. 1998. Simultaneous observations of tuna movements and their prey by sonic tracking and acoustic surveys. Hydrobiologia 371/372:61-69.

Josse, E., L. Dagorn, and A. Bertrand. 2000. Typology and behaviour of tuna aggregations around fish aggregating devices from acoustic surveys in French Polynesia. Aquat. Living Resour. 13:183-192.

Kleiber, P. and J. Hampton. 1994. Modeling effects of FADs and islands on movement of skipjack tuna (*Katsuwonus pelamis*): Estimating parameters from tagging data. Can. J. Fish. Aquat. Sci. 51:2642-2653.

Klimley, A. P., and C. F. Holloway. 1999. School fidelity and homing synchronicity of yellowfin tuna, *Thunnus albacares*. Mar. Biol. 133:307-317.

Lamkin, J. T., J. J. Govoni and T. D. Leming. 2001. Cold-core eddies, the Loop Current and larval tuna; a preferred spawning and nursery habitat? Abstract of paper presented at "Rigs and FADs" technical session of the Southern Division American Fisheries Society Meeting, Jacksonville, FL, January 22-25, 2001.

Marsac, F. and P. Cayre. 1998. Telemetry applied to behaviour analysis of yellowfin tuna (*Thunnus albacares*, Bonnaterre, 1788) movements in a network of fish aggregating devices. Hydrobiologia 371-372:155-171.

Martin, C. C. 1999. Results of the use of FADs in the pole and line skipjack (Katsuwonus pelamis) fishery in the North-West coast of Cuba. Pp. 38-39 in: J.-Y. Le Gall, P. Cayre and M. Taquet (Eds.), International Symposium on Tuna Fishing and Fish Aggregating Devices Book of Abstracts. IFREMER, Plouzane.

National Marine Fisheries Service (NMFS). 1997. 1994/1995 Report of the Southeast Fisheries Science Center Billfish Program. NOAA Tech. Memo. NMFS-SEFC-398.

Nemerson, D., S. Berkeley and C. Safina. 2000. Spawning site fidelity in Atlantic bluefin tuna, *Thunnus thynnus*: the use of size-frequency analysis to test for the presence of migrant east Atlantic bluefin tuna on Gulf of Mexico spawning grounds. Fish. Bull. 98:118-126.

Olander, D. 2001. Down the tubes: deep-dropping Gulf oil rigs. Sportfishing Magazine June 2001 <http://www.sportfishingmag.com/SF_Main/1,3214,5-1-1-23304-162-336,00.html>.

Ortiz, M. 2001. Review of tag-releases and recaptures for yellowfin tuna from the U.S. CTC Program. Col. Vol. Sci. Pap. ICCAT. 52:215-221.

Safina, C. 1993. Bluefin tuna in the West Atlantic: negligent management and the making of an endangered species. Conserv. Biol. 7:220-234.

Sakagawa, G. T. 1986. Skipjack fisheries in the western Atlantic. Pp. 99-103 in: P. E. K. Symons, P. M. Miyake and G. T. Sakagawa (Eds.), Proceedings of the ICCAT Conference on the International Skipjack Year Program. ICCAT, Madrid.

Sloan, R. 2001. Boom off Texas. Saltwater Sportsman Magazine June 2001:79-113.

Snyder, D. B., L. Lagera, P. Arnold, L. de Wit. G.H. Burgess and C. Friel. 2001. Potential interactions between pelagic longline fishing and deepwater oil gas structures in the Gulf of Mexico. Abstract of paper presented at "Rigs and FADs" special session of the American Fisheries Southern Division Meeting, January 22-25, 2001, Jacksonville, FL.

Wilson, C. A. 1988. Longlining for yellowfin tuna in the Gulf of Mexico. Louisiana Sea Grant College Program. 15 pp.

Deep Gulf Fisheries
Click on the title to view the presentation.

Overview of the Current Status of Socioeconomic Studies in the MMS Gulf of Mexico Region

Steve Murdock
Texas A&M University, College Station, Texas

Steve Murdock presented the preliminary findings of a White Paper, which he authored with F. Larry Leistritz and Stan Albrecht. The paper reviews MMS's socioeconomic studies program for the Gulf of Mexico and is intended to provide workshop participants with an overview of the work that has been conducted since the last deepwater workshop. The paper will be used to frame the socioeconomic breakout group discussions.

The White Paper focuses on recently completed studies and evaluates them using the following criteria:

1. The extent to which they address the key socioeconomic dimensions related to socioeconomic impact assessment (SIA);
2. The appropriateness, application, and documentation of the methodologies used;
3. The quality of these studies relative to conceptual, methodological and state-of-the-art criteria; and,
4. The likely utility of the work for communities and MMS managers and decision-makers.

The authors found the timing of this review to be unfortunate because much of the key research is ongoing, hence its findings cannot be evaluated. This means that the findings of the White Paper must remain preliminary.

The authors found that the MMS GOM socioeconomic studies program has shown substantial progress in the last decade. The agency has listened and responded to its critics and constituents and is aware of the critical socioeconomic dimensions in OCS development.

For example, in 1992 a GOM Socioeconomic workshop and the National Research Council recommended studies on historical OCS impacts, public attitudes toward industry, effects on families, worker attitudes, and education. This research has been completed.

The White Paper divides the review into three SIA dimensions: economic and fiscal, demographic and infrastructure, and sociocultural.

Economic and Fiscal

MMS GOM economic modeling has shown great progress over the last decade. After years of not being state-of-the-art, it is nearly state-of-the-art in all regards. MMS should continue building on these efforts. The White Paper emphasizes the continuing need for up-to-date information on industry expenditures and labor. Global competition and the industry's long-term role in the Gulf should also be considered.

Fiscal issues have not been as systematically addressed although studies of the Florida Panhandle and Port Fourchon do take state-of-the-art approaches. Similar case studies elsewhere would address this issue more fully.

Demographic and infrastructure

The evaluation of Gulf demographic research is premature because of critical ongoing studies. However, the Gulf's efforts in this area to date are not state-of-the-art. Developing more sophisticated demographic and public service model components should be given priority.

Major demographic issues that need to be addressed are in- and out-migration, population retention, wage pressures, and foreign immigration. A Gulf study is addressing industry support infrastructure. However public service components that are a standard part of SIA have not been as thoroughly addressed. Examples include health, welfare, public safety, power, water, sewer, transportation, and recreational facilities.

Sociocultural

In the Gulf, research in this area has shown great progress and is increasingly approaching state-of-the-art in terms of methods used and subjects addressed. Of particular note in this regard is the work on families.

Work should be extended to additional places and substantive areas and be expanded to identify a wider range of positive as well as negative effects. An effort should be made to more fully specify the groups that benefit and the ones that do not and to identify differences in areas and impacts across the Gulf.

The White Paper identifies several overall issues that should be considered when designing future research:

1. the difficulty in the separation of the baseline from OCS developmental effects;
2. the difficulty of identifying the impact area;
3. the problems of assessing commutative effects;
4. the need to more fully incorporate study findings into the EIS process; and,
5. the desire to make the EIS process more valuable to communities.

An Examination of Selected Recent MMS Socioeconomic Studies and Assessments in the Gulf of Mexico

Click on the title to view the presentation.

An Examination of Selected Recent MMS Socioeconomic
Studies and Assessments in the Gulf of Mexico

Draft Report for the Minerals Management Service
Gulf of Mexico Office

by

Steve H. Murdock
Department of Rural Sociology
Texas A&M University

F. Larry Leistritz
Department of Agricultural Economics
North Dakota State University

Stan Albrecht
Department of Sociology
Utah State University

May, 2002

V. Breakout Group Summaries

Ecology Breakout Group

Co-Chairs:
Michael Rex, University of Massachusetts
Robert Avent, Minerals Management Service
James P. Ray, Shell Global Solutions (US) Inc.
Jeff Childs, Minerals Management Service
Recorder:
Margaret Metcalf, Minerals Management Service

Introduction

The aims of the Ecology Breakout Group in the broader context of the 2002 Deepwater Workshop were to:

- review the current status of industry activity in deepwater
- discuss environmental issues that have emerged since the first deepwater workshop five years ago (Carney 1997)
- identify remaining data gaps and useful syntheses of information, relevant to points one and two above, that might form the basis of future environmental studies.

The first two objectives were covered largely through presentations by petroleum industry representatives and selected principal investigators of ongoing MMS-sponsored studies during the first day of the workshop. The Ecology Breakout Group focused its attention on reviewing the current status of environmental knowledge about the deep Gulf, and discussing what new initiatives might address environmental impacts of deepwater petroleum exploration.

The Ecological Issues Working Group of the first Deepwater Workshop (Carney 1997) reviewed deep-sea benthic ecology, including previous research in the Gulf of Mexico. The Working Group recommended a more comprehensive study of the structure and function of the deep benthic ecosystem, and intensified research on the ecology of recently discovered chemosynthetic communities begun in 1990 through the MMS-funded "Northern Gulf of Mexico Chemosynthetic Ecosystem Study" (CHEMO I). A continuation of the latter program, "Change and Stability in Gulf of Mexico Chemosynthetic Communities" (CHEMO II), has now been completed. The final report has been reviewed, and has recently been released. In response to recommendations of the first Deepwater Workshop, MMS sponsored the "Deepwater Program: Northern Gulf of Mexico Continental Slope Habitats and Benthic Ecology" (Deep Gulf of Mexico Benthos, DGoMB) study which began in 1999 and is still in progress. The study was extended into Mexican waters of the southern Gulf this year. The DGoMB study is intended to provide a much more thorough biotic survey of the deep Gulf, and to validate a model of ecosystem function in terms of carbon flux. Important conceptual advances of the DGoMB study are that it incorporates a sampling design to test specific hypotheses of community structure formulated from results of earlier dredging, and that the process-oriented component is an entirely new direction that integrates energy flow in the benthic community. Discussions of the 2002 Ecology Breakout Group concerning the present state of knowledge were based on what might realistically be anticipated from the CHEMO II study (not officially released when we met) and the DGoMB study which is ongoing. Clearly, priorities for future studies must take into account the final results of CHEMO II and DGoMB when they become available.

It is clear that petroleum exploration and production have extended into the deep Gulf far ahead of assessment studies to evaluate their potential environmental impacts (Baud et al. 2002). Also, physical and biological oceanographic research in the deep Gulf during the last decade has revealed an extremely complex ecosystem that includes many unanticipated novel habitats and ecological phenomena. Information is limited or not available on the geographic distribution of these habitats, the structure and function of their biological communities, and the ecological coupling among habitats, especially in deep water. Thus, designing new studies that can measure the environmental effects of deepwater drilling in a direct way presents a considerable challenge.

Following an initial discussion of general environmental problems associated with deepwater drilling, the Ecological Breakout Group formed two subgroups, Benthic Systems (Chaired by M. Rex, R. Avent, and J. Ray) and Pelagic Systems (chaired by J. Childs). Later we reassembled to review our independent conclusions and to look for ecological patterns and environmental concerns common to pelagic and benthic realms. Below we present the substance of these discussions and conclusions, and make several recommendations about which there appeared to be a broad consensus among environmental scientists, industry representatives and government regulators.

Benthic Systems Session

In the last several decades, ecological studies in the deep Gulf have progressed from descriptive biotic surveys to much more advanced approaches involving sampling programs designed to test biogeographic hypotheses, manipulative experiments and theoretical modeling. This is an impressive record and the pace of discovery has been dramatic. A more comprehensive picture of ecosystem structure and function is beginning to emerge, at least in broad outline. Discussions of the 2002 Ecology Breakout Group were concerned with building on this knowledge base to orient prospective studies more toward the immediate environmental effects of deepwater drilling.

The Benthic Session discussed five basic issues that are relevant to determining the environmental effects of deepwater drilling.

1. *Spatial and Temporal Scale.* Pattern and scale constitute the central problem of modern ecology (Levin 1992); a clear recognition of this is imperative to any applied study. Have we identified the relevant spatial and temporal scales at which environmental factors act to structure deep Gulf benthic communities? Deep-sea communities show geographic variation (Rex et al. 1997) that is shaped by processes operating at local (Grassle and Maciolek 1992),

regional (Cosson-Sarradin et al. 1998) and global (Rex et al. 1993, 2000) scales. Moreover, there is a complex interaction among phenomena occurring at different scales (Levin et al. 2001). For example, local species diversity and community makeup represent a balance between local ecological opportunity and continued dispersal from the regional species pool. Similarly, the ecology of local deep-sea communities varies on time scales ranging from annual (Gage and Tyler 1991) and decadal (Smith and Kaufmann 1999; Smith et al. 2001) to millennial (Raymo et al. 1998) and longer (Cronin and Raymo 1997). Predicting the impact of disturbance by drilling and the recovery potential of deep-sea communities depends on understanding the spatio-temporal scales on which ecological structuring agents function.

2. *Distinguishing Disturbance by Drilling from Natural Cycles.* The main problem in environmental impact studies is to separate natural spatio-temporal changes from those caused by anthropogenic activities. As pointed out above, deep-sea communities do show spatial variation on local to large scales, and experience temporal changes on short- and long-term scales. To assess the effects of deepwater drilling, it is necessary to know the type and magnitude of environmental change caused by drilling operations, the spatio-temporal variability of affected organisms and how these organisms respond to drilling activity. Much of the relevant basic information about community structure and dynamics may emerge from the DGoMB study. It is possible to design a controlled precision sampling study to distinguish drilling impacts from natural variation. The sampling design and its associated statistical tests require sampling multiple impact and control sites before and after drilling is initiated (Bernstein and Zalinski 1983; Stewart-Oaten et al. 1986; Underwood 1996, 1997). This approach might be particularly timely since drilling activity is currently expanding into deep water and opportunities exist for carefully planned before and after sampling programs that capture the appropriate spatial and temporal scales of natural variation. Clearly, such studies need to be long-term and will require taxonomy at the species level. The level of sample replication required can be determined by a power analysis (Cohen 1977); again, the results of the DGoMB study will be very useful in this regard.

Oil industry representatives point out that the distribution of cuttings and associated mud in the deposition zone around drilling sites is heterogeneous. Because of this, a sampling program would have to be very precise, or the sampling strategy would need to include enough random stations to adequately characterize the physical, chemical and biological changes that have occurred in this impact zone. A similar density of sampling would have to be conducted at multiple reference areas so that local variability could be adequately defined.

Other approaches to detecting impacts should also be considered. It may be possible to include potential anthropogenic disturbances into the food-web model being developed in the DGoMB study. Controlled experiments are the only way to directly determine a causal connection between drilling activity and changes in the natural community. All of these potential studies require careful selection of the appropriate response variable (e.g., standing stock, carbon flux, species diversity, and community composition, local or global extinction), and consideration of what constitutes an acceptable level of disturbance to the natural ecosystem. We stress again that it would be premature to design and implement such studies without considering the final results of the historic NGOMCS and the recently concluded (CHEMO II) and ongoing (DGoMB) MMS studies.

Conducting any of these studies in deeper water will be difficult, primarily because of the density of organisms decreases rapidly with depth. Lower animal abundance (and possibly the slower response times that attend colder and nutrient-poor environments) may mean that intensive sampling over a long period is necessary to detect both natural variation and potential impacts. Because of this, the Ecology Group urges a careful examination of existing information to optimize and inform the impact studies, including archival material and databases from earlier studies. Coordination of MMS, industry and environmental scientists is especially important. For example, industry representatives in the Ecology Group suggested that ROVs used on deepwater rigs may be available for sampling and conducting experiments. There was a general agreement among participants that the ability to distinguish drilling disturbance from natural cycles is a very high priority. MMS might consider a special workshop to explore existing data and to design studies using different possible approaches (sampling, experiments, modeling) to measure the potential impact of deepwater drilling in the Gulf.

3. *Understanding Biodiversity at Relevant Levels of Organization.* The concept of conserving biodiversity is an appropriate and useful context in which to consider the environmental effects of deepwater petroleum exploration. It has become the major focus of national and international efforts to preserve and manage natural ecosystems, and to plan for sustainable development. The industry perspectives presented at the Deepwater Workshop by representatives from British Petroleum and ChevronTexaco stressed the significance of biodiversity and minimizing potential impacts to biological communities.

In general, biodiversity simply means the variety of living organisms. It involves all levels of organization including landscape (or perhaps more appropriately "seascape"), community-ecosystem, population-species and genetic (Lubchenco et al. 1991; Norse 1993; NRC 1995; Heywood 1995; Meffe and Carroll 1997; Frankham et al. 2002). The Gulf shows extraordinary biodiversity at the landscape level (variety of habitats) including numerous soft-sediment conditions, chemosynthetic systems, gas hydrates, brine pools, *Lophelia* colonies, oxygen minimum zones, iron stone pavements and a vast array of unique topographic features. These habitats are ecologically linked. Exploiting a variety of habitats, directly or indirectly, may be vital to species persistence. For example, some mobile benthic predators derive their nutritional needs from both photosynthetic and chemosynthetic production (MacAvoy et al. 2002).

The DGoMB study will produce a great deal of information about the population-species and the community-ecosystem levels of organization. Geographic patterns of species diversity and faunal turnover, relative abundance distributions of species, life-history attributes, and food-web structure can reveal a great deal about the susceptibility of populations and communities to disturbance. Mapping species ranges is also fundamentally important for estimating the likelihood of extinction from environmental impacts. It is the primary evidence for predicting whether an impact on some scale of time and space will cause local or complete extinction. Species with sufficiently large geographic ranges may suffer local extinction near the impact site, but subsequently be able to recolonize from elsewhere in their ranges. Species with small ranges situated near the impact zone are more vulnerable to permanent extirpation. Hence, the degree of endemism (taxa restricted to a geographic area) in impacted areas is very relevant to extinction potential. Species' geographic ranges are poorly known in any deep-sea environment. The combination of DGoMB data, earlier databases and archived material could contribute very substantially to mapping species' ranges in the Gulf.

Very little is known about biodiversity at the genetic level in deep-sea species. Techniques are now readily available for sequencing DNA of marine species, even from preserved specimens. This can be done in a very cost effective way, since the material is already being collected in MMS studies for other purposes. Genetic population structure is a critical component of biodiversity for conservation efforts in coastal and terrestrial systems (Frankham et al. 2002). Habitat loss and reduced population size erode genetic variation making populations more vulnerable to extinction and less able to adapt to changing environmental circumstances. Local extinction can, in turn, destabilize local communities and lead to lower species diversity. Genetic population structure can also reveal the extent and direction of migration (gene flow) which reflect the role of dispersal in maintaining population viability and community structure. Because of the importance of genetic biodiversity studies to resolving basic problems of vulnerability to extinction within species and the causes of benthic community structure, MMS might consider a pilot project using abundant widely distributed species to evaluate the usefulness of this approach for impact assessment.

4. *Interdisciplinary Approaches.* MMS-sponsored studies in the Gulf have hugely increased our understanding of physical oceanography, marine geology, fisheries, deep-sea ecology as well as social and economic factors. It is now possible to forge a much more comprehensive multidisciplinary understanding of the Gulf ecosystem. For example, new information on current systems has an important bearing on larval dispersal and deep-sea community structure. High-resolution sonar imagery of bottom topography makes it possible, for the first time, for ecologists to ask specific questions about large-scale patterns of community structure and landscape-level biodiversity. In future Deepwater Workshops, we encourage interaction among the breakout groups to integrate and synthesize information from MMS-supported studies in geochemical, physical, biological and social disciplines, and from the oil industry.

5. *Importance of Integrated Databases.* Much of the data resulting from earlier MMS studies remain unpublished and difficult to access. The data and archived material are very valuable resources for designing specific studies to measure the impact of deepwater drilling. We encourage MMS to consider establishing a professionally managed relational database of earlier (and current) studies that is available on the internet to industry, regulators, environmental scientists and the public. The database should also specify the location of archived material. This would greatly increase the benefits of these studies for both applied and basic research. Making the data available to the broader community would also greatly expand the potential for integration and synthesis among these databases, and with other deep-sea studies. While the regulatory responsibilities of MMS concern applied work related to offshore petroleum exploration, much information of very basic scientific interest emerges from these studies. The studies have also suggested important basic research not directly

related to the MMS mandate. We urge MMS to integrate their priorities and programs with those of other funding agencies that support marine research.

The Ecology Breakout Group also discussed possible future studies, relevant to deepwater drilling, in specific Gulf habitats. For the sake of convenience, we divided habitats into chemosynthetic and nonchemosynthetic.

1. *Chemosynthetic Communities.* The current status of knowledge on chemosynthetic habitats was difficult to evaluate in a thorough way because the CHEMO II final report was not released, and no principal investigators from this study participated in our session. Greg Boland and Robert Avent of MMS, pointed out that deeper seep systems in the Gulf were largely unexplored because most available submersibles used to study them had depth ranges limited to 1000 m. Since drilling has already extended much deeper than 1000 m, future investigations of the distribution, function and structure of seep communities below 1000 m is an obvious priority. One suggested approach to compare the recovery potential of seep communities above and below 1000 m is to conduct experiments on recruitment and growth rates of constituent species. The methodologies for doing this are well established. Oil industry representatives made the point that seeps are now protected and that drilling activity is not allowed within a certain distance of known seeps.

2. *Non-chemosynthetic Communities.* While chemosynthetic communities are unique and important, non-chemosynthetic communities dominate the Gulf seafloor. As already mentioned, the DGoMB study will produce a wealth of information about the structure and function of soft sediment communities in the deep Gulf. This information will be essential for designing experiments and precision sampling studies and constructing models to measure the effects of deepwater drilling in a more direct way. It is clear that long-term datasets and an ecosystem-wide understanding of pattern, process and scale are essential to assess impacts.

Most of our discussion centered on recently discovered habitats and ecological coupling among these habitats. William Schroeder briefed us on *Lophelia* (non-reef building coral) colonies, which may be widespread in the deep Gulf. They remain poorly know in terms of their biology and distribution, and have become an international priority for conservation. Other novel Gulf habitats include iron stone pavements, brine pools, gas hydrates, carbonate outcrops, oxygen minimum zones, and a huge range of topographic features including small bathyal basins, furrows, knolls and escarpments. Shipwrecks may provide information about patterns of ecological succession in hard-substrate communities and provide valuable information about whether or not they afford value as deepwater artificial reefs. Industry is interested in this issue, because future decisions will need to be made on whether or not deepwater subsea equipment will need to be, or should be, removed at the end of the oil fields lifespan.

It is important to recognize, as already shown with seep and surrounding soft-sediment communities, that the rich variety of deep Gulf habitats collectively form the landscape level of biodiversity, and not a group of isolated entities. The ecological linkage between habitats, for nutrient availability and species range maintenance, is vital to the persistence of species, and hence communities. We need to know much more about these unique habitats and their functional role in landscape biodiversity in order to predict the effects of disturbance from deepwater drilling.

Recommendations of the Benthic Systems Session

- Design studies to distinguish the disturbance by drilling from natural cycles.
- Extend investigations to chemosynthetic systems below 1000 m.
- Improve our understanding of ecosystem function and biodiversity at the landscape level.
- Update plans for future studies after receipt of the DGoMB final report.

References

Baud, R. D., R. H. Peterson, G. E. Richardson, L. S. French, J. Regg, T. Montgomery, T. S. Williams, C. Doyle and M. Dorner. 2002. Deepwater Gulf of Mexico 2002: America's expanding frontier. OCS Report MMS 2002-021, U.S. Department of the Interior, Minerals Management Service, Gulf of Mexico OCS Region, New Orleans, Louisiana, 133 pp.

Bernstein, B. B. and Zalinski. 1983. An optimum sampling design and power test for environmental biologists. Journal of Environmental Management 16:35-43.

Carney, R. S. 1997. Workshop on environmental issues surrounding deepwater oil and gas development: Final report. OCS Study MMS 98-0022. U.S. Department of the Interior, Minerals Management Service, Gulf of Mexico OCS Region, New Orleans, Louisiana, 163 pp.

Cohen, J. 1977. Statistical power analysis for the behavioral sciences. Academic Press, New York.

Cosson-Sarradin, N., M. Sibuet, G. L. J. Paterson and A. Vangriesheim. 1998. Polychaete diversity at tropical Atlantic deep-sea sites: environmental effects. Marine Ecology Progress Series 165:173-185.

Cronin, T. M. and M. E. Raymo. 1997. Orbital forcing of deep-sea benthic species diversity. Nature 385:624-627.

Frankham, R., J. D. Bollou and D. A. Briscoe. 2002. Introduction to conservation genetics. Cambridge University Press, Cambridge.

Gage, J. D. and P. A. Tyler. 1991. Deep-sea biology. Cambridge University Press, Cambridge.

Grassle, J. F. and N. J. Maciolek. 1992. Deep-sea species richness: regional and local diversity estimated from quantitative bottom samples. American Naturalist 193:313-341.

Heywood, V. H. (Ed). 1995. Global biodiversity assessment. United Nations Environmental Programme. Cambridge University Press, Cambridge.

Levin, L. A., R. J. Etter, M. A. Rex, A. J. Gooday, C. R. Smith, J. Pineda, C. T. Stuart, R. R. Hessler and D. Pawson. 2001. Environmental influences on regional deep-sea species diversity. Annual Review of Ecology and Systematics 32:51-93.

Levin, S. A. 1992. The problem of pattern and scale in ecology. Ecology 73:1943-1967.

Lubchenco, J., A. M. Olson, L. B. Brubaker, S. R. Carpenter, M. M. Holland, S. P. Hubbell, S. A. Levin, J. A. MacMahon, P. A. Matson, J. M. Melillo, H. A. Mooney, C. H. Peterson, H. R. Pulliam, L. A. Real, P. J. Regal and P. G. Risser. 1991. The sustainable bioshpere initiative: an ecological research agenda. Ecology 72:371-412.

MacAvoy, S. E., R. S. Carney, C. R. Fisher and S. A. Macko. 2002. Use of chemosynthetic biomass by large, mobile, benthic predators in the Gulf of Mexico. Marine Ecology Progress Series 225:65-78.

Meffe, G. K. and C. R. Carroll. 1997. Principles of conservation biology. Sinauer, Sunderland, MA.

National Research Council - National Academy of Sciences (NRC). 1995. Understanding marine biodiversity. National Academy Press, Washington, D.C.

Norse, E. A. (Ed). 1993. Global Marine Biological Diversity: A strategy for building conservation into decision making. Island Press, Washington, D.C.

Raymo, M. E., S. Carter, D. W. Oppo and J. McManus. 1998. Millennial-scale climate instability during the early Pleistocene epoch. Nature 392:699-702.

Rex, M. A., R. J. Etter and C. T. Stuart. 1997. Large-scale patterns of species diversity in the deep-sea benthos. Pp 94-121, in: R. F. G. Ormond, J. D. Gage and M. V. Angel (Eds.), Marine biodiversity: patterns and processes. Cambridge University Press, Cambridge.

Rex, M. A., C. T. Stuart and G. Coyne. 2000. Latitudinal gradients of species richness in the deep-sea benthos of the North Atlantic. Proceedings of the National Academy of Sciences U.S.A. 97:4082-4085.

Rex, M. A., C. T. Stuart, R. R. Hessler, J. A. Allen, H. L. Sanders and G. D. F. Wilson. 1993. Global-scale latitudinal patterns of species diversity in the deep-sea benthos. Nature 365:636-639.

Smith, K. L., Jr. and R. S. Kaufmann. 1999. Long-term discrepancy between food supply and demand in the deep eastern North Pacific. Science 284:1174-1177.

Smith, K. L., Jr., R. S. Kaufmann, R. J. Baldwin and A. F. Carlucci. 2001. Pelagic-benthic coupling in the abyssal eastern North Pacific: an 8-year time-series study of food supply and demand. Limnology and Oceanography 46:543-556.

Stewart-Oaten, A., W. M. Murdoch and K. R. Parker. 1986. Environmental impact assessment: "pseudoreplication" in time? Ecology 67:939-940.

Underwood, A. J. 1996. On beyond BACI-sampling designs that might reliably detect environmental disturbances. Pp. 151-175, in: R. J. Schmitt and C. W. Osenberg (Eds.), Detecting Ecological Impacts. Academic Press, San Diego.

Underwood, A. J. 1997. Experiments in ecology: their logical design and interpretation using analysis of variance. Cambridge University Press, Cambridge.

Workshop on Deepwater Environmental Studies Strategy: Ecology Breakout Group
Click on the title to view the presentation.

**Workshop on Deepwater Environmental Studies Strategy:
A Five-Year Follow-Up and Planning for the Future**

Ecology Breakout Group

Co-Chairs:
Robert Avent, Minerals Management Service,
Michael Rex, University of Massachusetts
James P. Ray, Shell Global Solutions (US) Inc.
Jeff Childs, Minerals Management Service
Recorder:
Margaret Metcalf, Minerals Management Service

Pelagic Systems: Ecology Breakout Subgroup

Chair: Jeff Childs, Minerals Management Service

Humans have plied oceanic waters (greater than 200 m in depth) of the northern Gulf of Mexico for centuries, but, their presence at any one site has been ephemeral, lasting perhaps seconds to minutes. The advent of placing semi-permanent fixed and floating OCS structures in the oceanic waters of the northern Gulf introduces an ongoing human presence, unlike that experienced in the past. And, like other frontier areas that humans have settled and developed elsewhere, there will be environmental impacts to the oceanic areas of the northern Gulf resulting from the introduction of semi-permanent OCS structures and ensuing associated activities. These impacts may be greatest to pelagic organisms, which by definition includes the flora and fauna inhabiting the open waters of the ocean. Many organisms live part, if not all, of their existence in the water column. For example, pelagic fauna include, but are not limited to, sperm whales, sea birds, tuna, grouper, flounder, crabs, bryozoans, and corals. Some benthic organisms, such as many corals, which are regarded as sessile (i.e., stationary) organisms, may undergo a pelagic stage early in life, after being spawned by sessile parents. Coral "spat" may then settle on a substrate whereby they affix themselves and may eventually form a colony. For many species undergoing a pelagic phase in life, that phase frequently occurs when they are neonates and juveniles; life history stages that are most vulnerable to various impact vectors that can have sub-lethal and lethal effects.

The Pelagic Systems Session was tasked with identifying data gaps and useful syntheses of information relevant to the expansion of OCS Industry activities into oceanic waters of the northern Gulf of Mexico and potential environmental impacts associated with that expansion, so that such information might form the basis for future environmental studies. The session was attended by industry representatives, marine scientists, and government regulators, who identified basic and applied scientific information needs associated with the topic. The participants also ranked information needs based on their knowledge of existing information and opinions.

The basic information needs recognized in the Pelagic Systems Session primarily focus on identifying and understanding baseline environmental patterns and processes in the oceanic waters of the northern Gulf. Some information needs reinforce those identified in the Gulf of Mexico Deepwater Operations and Activities Environmental Assessment (MMS 2000-001).

Participants in the Pelagic Session emphasized the need for magnifying the scale of resolution concerning biotic distributions. For example, an atlas delimiting the three-dimensional distributions and abundances of each life history stage of keystone species would provide better information for management decisions. Studies to identify marine wildlife-habitat associations are very much needed. Such information, will improve our ability to assess the individual or cumulative impacts that semi-permanent fixed or floating OCS structures or associated activities may have on marine wildlife or their habitats. Specific information needs distinguished by participants of the Pelagic Systems Session include:

I. Three-dimensional hydrographic mapping of currents of the Gulf of Mexico
II. Production of an ecological atlas of the Gulf of Mexico to include:
   a. GIS mapping of biotic resource distributions and abundances
   b. Three-dimensional ecological mapping of living resources at various scales (e.g., community, assemblage, species, population, sex, life history stage) that include:
      i. delineation of biotope types
      ii. vertical & horizontal distributions and abundances
      iii. temporal (e.g., seasonal) distributions and abundances
      iv. scalar patterns of distribution relative to phys-oceanographic and bathymetric features (e.g., sperm whale cow and calf pairs relative to warm and cold core water masses)
III. Fish & wildlife-habitat association studies
   a. delineation of important habitat areas by use (e.g., mating, nursery, feeding, etc.)
   b. biotic (e.g., community, assemblages, species, populations, life history stages, sexes) vs. abiotic (e.g., physical-oceanographic and bathymetric) patterns and associations
IV. Pelagic-benthic linkages
   a. Energy pathway studies: to include investigations within oceanic provincial ecosystem, as well as between neritic and oceanic provinces
      i. Food web modeling
      ii. Stable isotope studies
V. Information search and synthesis of survey methodologies available for studying pelagic organisms
VI. Identification of / monitoring of keystone species or assemblages representative of the oceanic Gulf for assessing changes in the ecology of the region.
   a. Bluefin / yellowfin tuna
   b. Sperm whale / beaked whales / Bryde's whale
   c. Leatherback sea turtle
   d. Squid
   e. Grouper
   f. Billfishes
   g. Oceanic Sharks
   h. Mesopelagic crustaceans

Applied information needs requiring scientific study include a diverse range of topics. Chief among these is the potential for semi-permanent fixed or floating OCS structures to function as artificial islands to Caribbean species that have yet to colonize continental shelf waters of the northern Gulf of Mexico. The Loop Current may transport non-indigenous species from the Caribbean Sea or Yucatan Peninsula to OCS structures placed in oceanic waters of the northern Gulf which may then be utilized by the non-indigenous species for dispersal into continental shelf waters of the northern Gulf. Such introductions may have a profound and significant impact on endemic populations, species, assemblages, or communities. Also important, are studies to investigate the potential effects that these structures may have upon the behavior of highly migratory species of fishes, birds, sea turtles, and marine mammals in the region. For example, will the placement of semi-permanent OCS structures in oceanic waters modify migratory behavior or movements of these creatures? The effects of anthropogenic sources of noise continue to be a growing concern with respect to marine mammals and sea turtles; such studies were previously described in the Gulf of Mexico Deepwater Operations and Activities Environmental Assessment (MMS 2000-001). Applied information needs, as identified and ranked by order of importance by participants of the Pelagic System Session are:

Positive / negative impacts of oil & gas structures / activities in oceanic province
- Species introductions / re-distributions
    - non-oceanic species (e.g., reef creatures)
    - Caribbean species not occurring in northern GOM
- Recruitment / survivorship
    - Sargassum assemblages including sea turtle hatchlings and juvenile fishes
    - planktonic larvae
    - "Sanctuary effect"
    - New habitat and niche space
- Modification of behavioral ecology
    - Movement / migration
        1. Fish Aggregating Devices: Migratory distractions?
        2. seabirds and neo-tropical migratory birds
        3. marine mammals & sea turtles
- Noise
    - noise measurements of:
        1. ambient and biogenic sources
        2. subsea structures
        3. floating structures
        4. catenaries / anchoring systems
        5. pipelines
        6. service vessel traffic
        7. helicopter traffic
- Fisheries
    - commercial & recreational fisheries
    - socioeconomic aspects
    - stocks
    - stock assessments
- Biotechnology
    - use of biofouling organisms for medical / commercial uses
- Physical impacts
    - entanglement of large whales in catenaries or anchoring systems
    - vessel strikes
- Fate & effects of hydrocarbon spills to pelagic species and the oceanic ecosystem
- Structure removals
    - Impacts from methodologies (explosive / non-explosive)
    - Removal vs. leave-in-place

Participants of the Pelagic Systems Session also identified some general recommendations relevant to the development of environmental studies stemming from the Workshop. Specifically, these involve:

I. Investigate what other agencies / entities are doing to investigate these topics (NOAA/NMFS/NOS/USGS/US Navy/NSF/ etc.)

II. Collaborate with other agencies / entities to pool resources and funding; this should include collaboration with other countries bordering the GOM (e.g., Mexico and Cuba).

III. Make data (not just reports) available to all…agencies, academics, etc.

Presentation Not Available

Geology/Geohazards Breakout Group

Co-Chairs:
Bill Bryant, Texas A&M University
Sarah Tsoflias, Minerals Management Service
Jim Niemann, ChevronTexaco
Recorder:
T. J. Broussard, Minerals Management Service

Geohazards of the Continental Slope and Rise, Northern Gulf of Mexico

The engineering and geological constraints on the continental slope off Texas and Louisiana related to hydrocarbon recovery will require both novel geological and geophysical surveys and engineering methods. Significant seafloor engineering problems in deep water include slope instabilities, both short-term (slump) and long-term (creep); pipeline spanning problems; mass transport from unknown causes; and unusual stiffness and strength conditions (Dunlap 97). The geohazards (engineering and geologic constraints) present in and on the central and western continental slope and rise are many in number and are mainly due to the activity of salt and rapid sedimentation. Specific examples include: faults (sediment tectonics, halokinesis); slope stability (slope steepening, slumps, creep, debris flow); gassy sediments (sediment strength reduction, hydrates, sediment liquefaction); fluid and gas expulsion features; diapiric structures (salt, mud, hydrates); seafloor depressions (blowouts, pockmarks, seeps); seafloor features (sediment waves, differential channel fill, brine-low channels, seabed furrows); shallow water flow; and deep water high-velocity currents (mega-furrows, seabed erosion).

Information "Needs" Identified at the 1997 Workshop and Progress

Mass Movement Processes: The most impressive research on mass movement processes is the extensive investigation of the mass movement processes that had taken place in the British Petroleum Mad Dog and Atlantis prospects. The first use of a deep-water AUV to obtain side scan sonar, multibeam and subbottom profiles and extensive coring documented extensively the mass movement processes active along portions of the Sigsbee Escarpment. These studies will be the benchmarks for future investigations.

Gas Hydrates: Knowledge of Gas Hydrate in the Gulf of Mexico has increased, but questions remain. The various DOE projects are well under way and several JIP's have been announced. A major coring cruise of the *R/V Marion Dufrense* in June 2002 recovered 30 piston cores up to 44 meters in length in areas of known hydrate presence. Observations using the *DSV NR-1, DSV Alvin* and *DSV Johnson SeaLink* were obtained in hydrate rich environments. Gas seeps were identified from satellite observations. Recent observations indicate that not many hydrate sites are found in ultra-deepwater GOM and raises the question is Shallow Gas an issue in deepwater?

Sediment Types: No assertive MMS program or project has addressed the issue of sediment types on the continental slope and rise of the Gulf of Mexico. The DGOMB Project supported by MMS has sampled several hundred box-core sites on the continental slope; Mississippi Fan and the Sigsbee Abyssal Plain have added information on the surficial sediment distribution in the Gulf of Mexico. Individual projects at Rice University, Texas A&M University, and the Universities of Texas and Louisiana have also pursued this issue.

Additional Concerns

Bathymetry: The use of 3D seismic to determine bathymetry is the only effort outside of general geohazard surveys to add to our knowledge of the continental slope and rise in the northern Gulf of Mexico.

Instrumentation: Technological and instrumentation improvements have taken great strides in the fields of ROV and AUV usage.

New Issues in Past 5 Years

Discovery of Mega-furrows and Intense Deepwater Currents

While collecting data during a deep-tow survey, which consisted of a 3.5 kHz subbottom profiler and a 100 kHz side-scan-sonar, of the northern Gulf of Mexico continental slope and rise around Bryant Canyon in the spring of 1999, a series of longitudinal mega-bed-forms were observed in ~3000 m of water along strike at the base of the Sigsbee Escarpment. The mega-bedforms surveyed included a field of sedimentary mega-furrows ~10 m deep and ~30 m wide, spaced ~100 m apart in a field 10-25 km wide, south of the escarpment. Earlier deep tow surveys in 1997 observed sedimentary furrows proximal to the escarpment north and south of Green Knoll in water depths ranging from 1400 m to 1900 m and on the slopes of Vaca Basin. Sedimentary furrows are longitudinal bed-forms that occur in fine-grained sediment, and have been observed in a wide variety of settings ranging from the deep ocean, to deep lakes, to the Hudson River. The existence of furrows in deep sea settings suggest that unidirectional bottom currents exist in the range of 1 to 100 cm/sec and higher. Such currents may be continuous or highly episodic, and the furrows are inferred to be formed due to helical secondary circulation. Sedimentary furrows were first produced in the lab by Allen (1969). Allen found a hierarchy of bed-forms ranging from uniform furrows at low flow conditions, to meandering furrows and flute casts; at the highest flow conditions, he found sheet flow. All the conditions described by Allen are present in the area due south of Bryant Canyon on the continental rise. The significance of these features is the high water velocities necessary for their formation and the large amount of erosion occurring at the base of the Sigsbee Escarpment. Indications are that the erosional processes are active during high-sea stands and depositional and infilling of the furrows are active during low-sea stands. The impact of these features and the currents on pipeline and platform activities may be significant.

An extensive field of mega-furrows has been recently discovered on the seafloor at the base of the Sigsbee Escarpment in the northwestern Gulf of Mexico (Bryant et al. 2000). Texas A&M deep-tow data and 3-D seismic data supplied by WesternGeco show that the scale of this field far exceeds anything previously observed in the world's oceans. These data allow the furrows to be resolved in unprecedented detail. The size of the furrows, variations in their morphology, and their orientation relative to large topographic features all suggest that the furrows are produced by strong deepwater flow. Individual furrows can be traced continuously for at least 50 km on the 3-D surface renderings, they are among the longest continuous individual sedimentary bed forms on the surface of the Earth. Data from a near bottom current meter confirms that strong, previously unknown, currents do exist at the base of the escarpment (Hamilton and Lugo-Fernandez 2001). The furrows and associated environmental processes are dominant features of the northwestern Gulf of Mexico and similar features will probably be found to exist in many other areas of the world's oceans.

Current and Future Deepwater Geology and Geohazard Information Needs as of July 2002

High Priority Items

1. Slope Stability / Pore Pressure
2. Sea Floor Morphology
3. Mega-furrows and Associated Currents

Slope Stability/Pore Pressure

MMS Significance: Need for these studies may grow as more deepwater activities face slope stability related problems. The complex bathymetry of the upper, middle and lower continental slope off Texas and Louisiana related to the halokinesis of allochthonous salt renders many intra-slope basins with high-angle basin walls. Slumps and slides are known to have occurred in many of the continental intraslope basins. The areas within and around the Mississippi Canyon have experienced extensive slope failure. The most important parameters in slope failure are the slope angle, the shear strength of the sediment and the pore pressure within the sediment. We know the general slope angles of the intraslope basin walls from multi-beam data but know very little of the shear strength and pore pressures within the continental slope and rise and along the Sigsbee Escarpment.

Impacts of slope stability on seafloor structures such as pipelines, seafloor templates, BLPs, etc is obvious, slumps and slides are extremely powerful events and seafloor structures can rarely be designed to withstand such forces. The best approach is locating structures in areas that will be least affected by such events.

Some of the questions that arise: is the data on hand relative to slumping relevant to deep-water operations? Are most of the observed slump features older that 2,000 to 10,000 year ago? Are slope failures a concern given the deepwater operational timeframe (~100 yrs)? What are the affects of high velocity bottom currents on the stability of the Sigsbee Escarpment?

To partially answer some of these questions we need to examine the knowledge of previous investigations/data collections such as was taken by BP for their Mad Dog and Atlantis Prospects. These two studies set the benchmark for future slope investigations.

A forum on "RISK ASSESSMENT FOR SUBMARINE SLOPE STABILITY" was held in Houston, Texas on May 10 and 11, 2002 and was sponsored by the Minerals Management Service. The following is part of the summary of the report following the meeting.

The Forum brought together a number of participants with broad views and backgrounds related to issues of risk assessment for slope stability in deep water. In reality the field is even much broader than represented by the participants. Views ranged from relatively narrow, well-focused views of technological needs in specific areas to broader views of issues such as integration and planning. Based on the discussion over the course of the Forum at least several observations were made:

- There is a need for better integration of geological, geotechnical, and geophysical data gathering and interpretation. The high cost of obtaining these data limits how much can be obtained and requires that the information be integrated to optimize its usefulness.
- There is a need for more advanced planning and investigation. "Paper" or "desktop" studies and studies in the very early stages of planning should improve the understanding and maximize what can be learned. Often there is neither time nor money to do as thorough an investigation as might otherwise be possible.
- The understanding of properties and processes at a single, site-specific point is very good compared to the understanding of properties and processes both spatially and temporarily. While additional resources might improve on this, better integration of data and earlier starts to investigations could also help.
- There is still need for improved technology and more data collection. How pore pressures develop and how soils behave as failure progresses, e.g., a change of state from a solid to a fluid, are examples of where further investigation is warranted. There is also need for fundamental understanding of the triggering mechanisms for slope failure and how that slope failure progresses.
- Standards comparable to those that have been developed for foundations and for risk assessment for earthquakes would be beneficial and probably could be developed even with the present state of knowledge.
- There is a lack of well-documented case histories. Often the triggering mechanisms are unknown and there is usually a paucity of good geological, geotechnical or geophysical data. In some instances well-documented cases may exist, but are not in the public domain and thus, not available to many.

Sea Floor Morphology

MMS Significance: Detailed Bathymetry Data is an underlying need for several Deepwater Research areas (i.e., currents, biological interactions); would assist preliminary operations planning.

There are large gaps in Western and Eastern Gulf of Mexico bathymetry data that generally include areas outside the NOAA multibeam data set taken a decade ago. The NOAA multibean bathymetry is the best data on hand that covers a good portion of the continental slope off Texas and Louisiana.

The highest resolution bathymetry available, besides AUV and Deep-Tow data sets, is the use of surface extraction of 3D seismic data sets taken by the seismic industry. The need for information sharing of bathymetry data extrapolated from existing 3D surveys would help to determine basic seafloor morphology, planning with regards to pipeline installation and routing and slope stability analysis and many other benefits. The resolution of the 3D bathymetry is an order of magnitude better than the existing NOAA multibeam data. Attempts should be undertaken to investigate how the seismic industry could benefit thru the release of 3D bathymetry data sets.

Mega-Furrows and Associated Currents

The Mississippi Fan Fold Belt/Green Canyon area of the Sigsbee Escarpment is the primary area of ultra-deepwater oil and gas prospects in the United States. In this area and the rest of the 500 km long escarpment as well, an extremely large field of mega-furrows on the seafloor and associated strong bottom currents have recently been discovered. Existing seismic data is limited, but initial analysis of it indicates that mega-furrows are long, relatively narrow depressions in the seafloor that are generally oriented parallel to the escarpment. Individual mega-furrows are ~1 to 10 m deep and ~5 to 50 m wide with ~20 to 200 m between individual furrows. The whole field of furrows appears to cover an area in excess of 15,000 km^2. Extremely limited current velocity measurements show that near bottom currents can exceed 2 knots (100 cm/sec). The spatial variations of furrow topography have not been characterized and the spatial and temporal variations of the bottom water flow field are also unknown. Knowledge of the spatial and temporal variations of furrows is necessary to understand the processes forming furrows and to predict the variations of sediment properties with depth below the seafloor. It will also provide important insight into the geological development of the continental rise.

There are significant practical reasons to better understand the mega-furrows and associated strong currents along the base of the escarpment and in other areas where strong currents erode muddy sediments. The following are several technical and environmental concerns associated with the furrows and associated currents:

- Extreme topographic relief and strong flow affects the pattern that cuttings and drilling mud might be deposited on the seafloor,
- Topographic relief and strong flow complicates the design and installation of pipelines,
- Erosional scour affects stability of pipelines, foundation piles, and seafloor installations,
- Slope undercutting by erosional scour potentially leads to hazardous slumps and slides,
- Strong bottom-water flow may produce vortex-induced vibrations that cause fatigue and failure of riser strings, platform tension legs, and anchor cables,
- In several recent instances, strong currents prevented the use of industrial type remotely operated vehicles (ROVs) intended to assess geohazards and inspect seafloor installations,
- Observational data derived from study of the furrows and associated strong bottom water flow will provide valuable data for calibration and validation of numerical models used to predict trajectories of pollutant plumes,
- To better understand the significance of channelized density flows related to the leaching of exposed salt along the Sigsbee Escarpment.

The Geology/Geohazards Breakout Group addressed several additional topics of concern including:

1. groundtruthing geophysical signatures indicative of sensitive biological features,
2. knowledge of canyons and associated currents,
3. naturally occurring gas hydrates,
4. subsidence associated with petroleum extraction,
5. shallow water flows,
6. salt movement and rheology,
7. shallow gas hazards in the deepwater Gulf of Mexico (GOM).

Subsidence and salt movement were cited as potential future problems as development and production of hydrocarbon progresses throughout the deepwater GOM. A similar MMS Deepwater Workshop in 1997 discussed gas hydrates, gas hazards and shallow water flows but additional information, experience and improved processes have reduced the scope and uncertainty of these problems. The advance of technical instrumentation and seismic data has revealed potential new concerns as well as the opportunity to better characterize, evaluate and protect valuable economic and biological assets.

Groundtruthing Geophysical Signatures Indicative of Biological Features

The Geology/Hazards Subgroup suggests a study to correlate and compare seismic signatures and attributes with known ocean floor biological communities. This study would guide the MMS in developing effective and appropriate avoidance regulations, least disruptive pipeline right-of-ways and more effective focusing of marine biologic research. Subset structures such as wellheads, flow lines, manifolds in addition to anchors and mooring lines must be located away (500 feet) from suspected ocean bottom dwelling biological communities. These communities can be inferred from seismic bathymetry data and side scan sonar imagery of faults scarps, gas vents or oil seeps that may support a biological community. In addition seismic amplitude plots of the seafloor reflection may be indicative of gas charged sediment, hydrates or carbonate accumulation associated with biological activity.

Positive verification of biological activity inferred by acoustic signatures require expensive remote-operated vehicles (ROVs) or manned submersible dives to obtain visual identification or more disruptive piston core and grab sampling. A comprehensive study of seafloor biological communities correlated with their seismic characteristics including amplitude, structure, dip, coherence and other attributes would be extremely useful for groundtruthing seismic investigations elsewhere. Such a study would dramatically increase the value of 2D and 3D seismic surveys to map out seafloor biological communities without additional expense or disruption to the environment. Calibration of the vast amount of available 3D seismic data could afford marine researchers a tremendous opportunity to conduct regional ecological surveys across much of the U.S. deepwater Gulf of Mexico.

Canyons and Associated Currents

Detailed studies of canyons and their associated currents will lead to better understanding of upwelling, nutrient cycles and the links to marine biology and ecology. There is now an opportunity with prolific seafloor bathymetry data from regional speculative 3D seismic surveys and the increased interest and study of deepwater bottom currents associated with mega furrow structures to understand and map out the interactions of GOM continental slope canyons, their associated currents and the impact on marine ecology.

Naturally Occurring Gas Hydrates

Naturally occurring gas hydrates and associated gas trapped beneath hydrates pose a potential hazard to well bore stability while drilling and during production. These hydrates also represent a possible significant energy resource for the future. The 1997 MMS Report on Environmental Issues in Deepwater cited gas hydrates as a potential and largely unknown hazard to installations and operation in the deepwater environment. Gas hydrates are crystalline solids composed of gas molecules (usually methane) enclosed in a lattice of water molecules. These substances can only exist where there is a supply of hydrocarbon (gas) at extremely low temperatures or relatively low temperatures and high pressures. The GOM continental slope is ideal for hydrate formation with nearly freezing temperature, high pressure and potential sources of gas from bacterial generation and migration of thermogenic hydrocarbon from vents and seeps along faults and fault complexes.

Observations from coring studies and industry exploration drilling suggest very little volume of hydrate in the ultra-deepwater GOM sediments. Most documented hydrate occurrences on the continental slope are associated with hydrocarbon seeps and vents along fault scarps, at the fractured margins of tabular salt bodies and at the top of salt sheets. These hydrate occurrences appear to be mostly surface occurrences rather than thick layers, although sparse data from shallow logs and drilling returns does not definitively rule out the presence of pervasive hydrate in the subsurface. The scarcity of shallow gas sands (relative to the shelf) in deepwater sediments suggests a relatively small supply of gas in ultra-deepwater sediments. In addition, the presence of tabular salt bodies across a vast section of ultra-deepwater area precludes the migration of significant thermogenic hydrocarbons to the shallow sediments (except along faults and the margins of salt bodies) where temperatures are low enough for hydrates to form. Beyond the salt escarpment, hydrocarbon source beds are not believed to be sufficiently mature to generate significant gas to the shallow section.

In addition to the studies mentioned above, ChevronTexaco has been contracted by the DOE to conduct a Joint Industry Project (JIP) study of naturally occurring hydrates in the Gulf of Mexico. The goals of this JIP are to:

- Develop technology to characterize naturally occurring deepwater hydrates in the GOM,
- Understand how natural gas hydrates affect seafloor stability,
- Gather data to aid climate studies,
- Determine how gas hydrates might act as trapping mechanisms for shallow oil or gas reservoirs. (http://www.theenergyforum.com/hydrates_032002/WorkshopReport.pdf)

Gas hydrates remain a concern for the long-term stability of selective deepwater subsea facilities and operation. The relative scarcity of subsurface hydrate in the deepwater GOM, as seen by exploratory drilling, mitigates the concern that hydrates are a pervasive geohazard, a threat to global warming and casts doubt on the hope of hydrates as a significant future gas resource (for the Gulf of Mexico). Current research efforts and existing geohazard identification practices appear adequate at this point to address present and potential future hazards to deepwater GOM development.

Long Term Subsidence Associated with Petroleum Extraction

Production of hydrocarbons from shallow, thick, overpressured reservoirs could induce subsidence at the seafloor negatively impacting subsea production facilities, wellheads, producing well bores, and pipelines. The earth's stress field has three basic components: the overburden or total stress (caused by the overlying weight of sediment and water), the pore pressure, and the effective stress (portion of stress borne by the sediment framework). Effective stress is equivalent to the overburden minus the pore pressure. As oil and gas is produced from reservoirs, the reservoir pressure may drop significantly over time increasing the effective stress. Porosity is exponentially related to the effective stress (Terzaghi and Peck 1948), so as the effective stress increases, porosity decreases and the reservoir (along with some portion of the bounding sediment) will compact. This compaction deformation will be transmitted (to more or less degree) through the overlying overburden to the surface as subsidence. Several workers (Geertsma 1973; Martin and Serdengecti 1984) have shown that the magnitude of the subsidence is related to the depth of burial, the degree of pressure decline, the consolidation state of the sediment, and the ratio of the thickness of reservoir to its area extent or diameter.

Spectacular examples of subsidence due to oil production exist in California, most notably the Wilmington field near the city of Long Beach (http://www.ci.long-beach.ca.us/oil/subsidence.html). International examples include the Groningen gas field in the Netherlands, the Bolivar Coastal oil field in Venezuela and the Po Delta field in Italy (Xu et al. 2001). In the Gulf Coast, Pratt and Johnson (1926) documented about 1 meter of subsidence at the Goose Creek field near Galveston, Texas. Offshore, subsidence of >8 meters at the Ekofisk field in the North Sea has required an expensive program of initial platform jacking (6 meters) plus water and gas injection and extensive redevelopment (Balson et al. 2001). The Louisiana Division of Natural Resources (DNR) reviewed subsidence associated with oil and gas production in Louisiana and concluded that subsidence is probably widespread but so minimal that it is not likely to be noticed (http://www.dnr.state.la.us/SEC/EXECDIV/TECHASMT/lep/subsid/subsid.htm). The DNR cites Martin and Serdengecti's (1984) analysis of Louisiana oil and gas reservoirs that showed that most reservoirs in South Louisiana are deeply buried, relatively thin, have low thickness to areal extent ratios, and are normally (hydrostatically) pressured often with strong water drives (thus maintaining pressure through production decline). Reservoirs in North Louisiana tend to be shallowly buried but the reservoirs are well consolidated and do not compact as readily with pressure decrease. All these factors tend to reduce the amount of reservoir compaction associated with extraction or mute the subsidence response at the surface. Operators in California and other parts of the world have minimized and in some cases reversed the effects of subsidence by artificially maintaining or increasing pressures in the formation reservoirs through gas or water injection, restraining production to allow water driven mechanisms to keep up with hydrocarbon extraction or ceasing production altogether (Poland 1984).

Xu et al. (2001) modeled subsidence in the Lost Hills/Belridge oil field area of Southern California as a function of the depth, radius of the reservoir, Poisson's Ratio, pressure drop due to production, thickness of the reservoir and Young's Modulus. They found good agreement of the predicted subsidence from pressure decline in the field with the actual subsidence measured by satellite radar interferometry.

Hydrocarbon reservoirs in the deepwater GOM typically are overpressured and undercompacted (with high porosities) even at shallow depths due to the rapid loading of deepwater sediments in intraslope salt basins and the poor hydraulic conductivity of the sand poor sedimentary section. Many reservoirs may be at shallow depths relative to seafloor. Due to large reserve requirements for commercial deepwater development and the often restricted areal extent of deepwater sand distribution, deepwater producing reservoirs likely will have high thickness to areal extent ratios. Other reservoirs however reside at great depths below the seafloor or may be covered by layers of mobile salt that will prevent compaction from inducing subsidence at the seafloor.

It is clear that there is potential for subsidence associated with production from deepwater GOM reservoirs but the magnitude of the subsidence and its impact on facilities is unclear. The critical factors in surface subsidence appear to be well understood and the MMS may have sufficient information in its databases to assess the long term risk for subsidence in existing deepwater producing fields. Given the level of understanding of the problem and potential solutions, subsidence due to oil and gas production may become a problem for select producing fields; however, it is unlikely to be a constraint or barrier to development and production across the deepwater Gulf of Mexico.

Shallow Water Flows

Shallow water flows are flows of water, gas and/or sand from shallow buried, overpressured sands in a deepwater setting, Shallow, overpressured sands are common in the deepwater where isolated sand bodies are deposited and encased in sealing shales and rapidly loaded by overlying sediments (Alberty et al. 1997). Well bore flows generally are prevented by the circulation of dense, weighted (greater than seawater) drilling muds down through a large pipe (marine riser) extending from the drilling rig through the water column to the wellhead on the seafloor. If the pressure exerted by the mud column is greater than the pressure in the permeable formation, the well bore will not flow. However, the consolidation and fracture strength of shallow, deepwater sediments is not sufficient to hold the pressure of a weighted mud column from sea level without fracturing and allowing fluids to escape to the seafloor. Therefore the first 1500-3000' of section below the mudline must be drilled without a riser pipe using seawater as the drilling fluid and risking the occurrence of flows from overpressured sands. Flows can range from gas bubbles to minor wispy flows to major turbulent flows that eject tremendous quantities of sand, mud, gas and formation water.

Deepwater, shallow water flows are primarily a conservation issue with subsidiary environmental and safety concerns. A shallow water flow necessitates time consuming, expensive circulation of heavy drilling mud in the open hole to "kill" the flow. If the flow cannot be controlled or plugged, then the well bore may be abandoned and a new hole drilled. Even if the flow is stopped, significant sediment may be eroded from the borehole to create a large void down hole that cannot be cemented behind pipe. The flow may resume (now behind pipe) as the weighted mud is displaced by lighter, less dense cement requiring additional cementation, circulation of mudweight behind pipe, and redrilling of the hole. Alternately the well bore may buckle because of inadequate support from the sediment as additional pipe weight is added to the wellhead. This can lead to junking of the well before objectives are reached or make the well unusable for future completion and production operations. Additional consequences include the ejection of sediment, formation and drilling fluids on the seafloor.

The MMS has sponsored and participated in a number of forums devoted to the understanding and prevention of shallow water flow. In addition, the industry has gained experience in drilling and mitigating shallow water flows and the adverse consequences of those flows. The primary method for prevention of shallow water flow is detailed seismic stratigraphic evaluation from exploratory 3D and/or high resolution 2D or 3D seismic. Drill locations can be selected to avoid or minimize the penetration of sedimentary facies that have a high risk for shallow water flow. Where moderate to high shallow water flow potential sites cannot be avoided, dense drilling mud can be pumped down the drill string while drilling in a riserless mode with the mud returning to the wellhead and depositing on the seafloor. The combination of low seawater pressure gradient in the ocean column and the higher pressure gradient of the heavy drilling mud yields an effective down hole pressure gradient sufficiently high enough to prevent formation water flow but not high enough to break down the weak, shallow formation. Drilling mud can be circulated at the first sign of flow detected by annular pressure monitored in real time down the hole or visually from a ROV monitoring the wellhead. Drilling mud may also be circulated at a predetermined depth to preemptively avoid the possibility of shallow water flow.

Shallow water flow sands remain a cost and drilling risk concern for deepwater drilling and development; however the issue now is considered an engineering constraint rather than a barrier to deepwater development and production.

Salt Movement and Rheology

Salt movement is a concern for seafloor modification and initiation of slope failure impacting pipelines and sea floor facilities. Movement of salt may initiate fault movement causing deformation and possible shearing of well bores. Movement of salt bodies at or near the seafloor causes deformation and disruption to sediments and the seafloor that readily can be observed from seafloor bathymetry maps and seismic cross sections. The time scale and rate of movement of salt deforming the seafloor needs to be studied to determine what impact, if any, will be had on casing, flow lines, subsea structures and pipelines.

Other features related to salt movement and rheology may also impact drilling and development operations. Shale inclusions, sutures (sediment entrained between merged salt bodies) and subsalt deformation zones may exhibit abnormally high pore pressures and/or abnormally weak formation integrity resulting in increased drilling costs, time and possible abandonment of resource objectives.

There is a great deal of research on salt movement and rheology related to salt mining, basin modeling and petroleum extraction which will not be reviewed here. Very little work however has focused on the impact of salt movement for drilling and producing operations in the deepwater marine environment. Although most impacts to deepwater operations by salt movement may turn out to be engineering constraints rather than barriers, the long term impact of salt movement to subsalt oil field development and production remains to be determined.

Shallow Gas Hazards in the Ultra-Deepwater GOM

Shallow gas sands are a significant safety hazard on the shelf for bottom-founded drilling rigs and production platforms. The shallow hazard/geotechnical evaluation industry and the extensive geotechnical regulatory framework have developed largely in response to this particular hazard. In the deepwater, the utilization of floating drilling and production structures mitigates much of the safety concern associated with breaching of the seafloor; however shallow gas sands remains a concern for subsea blowouts and well control. Deepwater operators frequently have encountered shallow water flow sands (often with associated gas) but rarely have encountered shallow gas reservoirs. Partly this may be a function of industry avoidance of obvious seismic amplitude ("bright spots") anomalies in the shallow seismic section. It may also be a function of scarce gas in the shallow section either due to the blocking of thermogenic gas by tabular salt sheets mentioned previously or a lack of bacterial generation or activity within shallow sediments. The possible suppression of bacterial activity, specifically methane generation, by the high pressures and low temperatures of the ultra-deepwater may be a good subject for future research. While shallow gas hazards always will pose a concern for marine drilling operators, the lack of shallow gas occurrences suggest that shallow gas is not a big concern in the ultra-deepwater and/or deepwater operators and regulators are doing a good job of prevention and avoidance of shallow gas problems.

References

Alberty, M. W., M. E. Hafle, J. C. Minge and T. M. Byrd. 1997. Mechanisms of shallow water flows and drilling practices for intervention. Offshore Technology Conference, Houston, TX. OTC 8301, Pp. 241-247.

Allen, J. R. L. 1969. Erosional current marks of weakly cohesive mud beds. J. Sedimentary Petrology 39:607–623.

Balson, P., A. Butcher, R. Holmes, H. Johnson, M. Lewis and R. Musson. 2001. North Sea Geology – Strategic Environmental Assessment (SEA2) Technical Report TR_008, British Geological Survey. Pp. 14-15.

Bryant, W. R., T. Dellapenna, A. Silva, W. Dunlap and D. Bean. 2000. Mega-Furrows on the Continental Rise South of the Sigsbee Escarpment, Northwest Gulf of Mexico. AAPG Annual Meeting, New Orleans La.

Geertsma, J. 1973. Land Subsidence above compacting oil and gas reservoirs. Journal of Petroleum Technology. Pp. 734-744.

Hamilton, P. and A. Lugo-Fernandez. 2001. Observations of high speed deep currents in the northern Gulf of Mexico. Geophysical Research Letters 28:2867-2870.

Martin, J. C. and S. Serdengecti. 1984. Subsidence over oil and gas field. Review in Engineering Geology Volume 6, Geological Society of America. Pp. 23-34.

Poland, J. F. and Working Group. 1984. Review of methods to control or arrest subsidence, Pp. 127-130, in Guidebook to Studies of Land Subsidence Due to Ground-Water Withdrawal: UNESCO. (Out of print but can be accessed electronically at http://www. rcamnl.wr.usgs.gov/rgws/Unesco/).

Pratt, Wallace E., and Douglas W. Johnson. 1926. Local subsidence of the Goose Creek Oil Field. Journal of Geology 34(7, part 1): 577-590.

Terzaghi, K. and R. B. Peck. 1948. Soil mechanics in engineering practice. John Wiley and Sons, Inc., New York, 566 pp.

Xu, H., J. Dvorkin and A. Nur. 2001. Linking oil production to surface subsidence from satellite radar interferometry. Geophysical Research Letters 28(7):1307-1310.

http://www.indiana.edu/~lcg/ACTIVE/SaltTectonics/salttectonics.html -Indiana University Laboratory for computation dynamics.

http://www.geol.lsu.edu/nunn/#Salinity –Jeff Nunn website.

Workshop on Deepwater Environmental Studies Strategy: Geology/Geohazards Breakout Group

Click on the title to view the presentation.

Workshop on Deepwater Environmental Studies Strategy

Geology/Geohazards Breakout Group

Co-Chairs

Bill Bryant, Texas A&M
Sarah Tsoflias, MMS
Jim Niemann, ChevronTexaco

Recorder

T. J. Broussard

Socioeconomics Breakout Group

Co-Chairs:
Steve Murdock, Texas A&M University
Harry Luton, Minerals Management Service
Recorder:
Connie Landry, Minerals Management Service

A major theme of the socioeconomic breakout group discussion was the recognition of the significant maturation that has occurred in socioeconomic studies in the Gulf of Mexico Region (GOMR) over the last several years. Just five years previously, in 1997, the socioeconomic breakout group for the MMS Gulf of Mexico Region Deepwater Workshop expressed frustration at the lack of completed studies to review. Similar concerns had been expressed previously during other reviews and at other workshops. However, by 2002, the picture has changed significantly. The last several years have produced a plethora of studies addressing a broad range of critical socioeconomic issues. Major themes identified by earlier reviews and workshops as necessary areas for research in the GOMR have been or are being addressed by a regular stream of studies. Many of these studies are now published and available to the public; others will soon be completed. As a consequence, the white paper prepared for discussion at the 2002 workshop was able to review and evaluate a much larger number of completed studies that extend across an impressive range of research issues. The one concern raised by participants at the workshop was that of timing; while a number of studies were available for review and discussion, a number of others are in the final stages of completion, but were not available to review.

While it would have been nice to have these to include in the review, it was acknowledged that much of the growth of the Environmental Studies Program in the GOMR in recent years has occurred in the socioeconomic arena. In this sense, the workshop was a celebration of the substantial progress that has been and is now being made.

In the socioeconomic breakout sessions, the white paper was used as a reference point, not as a focus of analysis. As described above, it summarizes and critiques work completed to date, and identifies future studies that are needed. This was used as a point of departure for the workshop discussions. Many of the conclusions and recommendations that emerged from those discussions are consistent with findings of the white paper. However, a number of new issues were identified, as well. Throughout the sessions, it was acknowledged that many of the gaps in research identified in the white paper, and by workshop participants, are being addressed by studies currently underway.

Based on the review of the white paper and group interaction, the following general conclusions emerged from the 2002 workshop:

1. Industry studies have become an important part of the socioeconomic study agenda, and should continue to be a key component of the studies program. For example, international

labor market studies are particularly critical since GOMR oil and gas activity must compete with international labor demands. Moreover, it is no longer possible to depend heavily on a local labor market, as may have been the case earlier.

2. Economic studies remain central to the studies program. The studies reviewed in this area by the white paper were found to be quite strong; economic projections were particularly well-developed. Among the recommendations in this area were the following: a) the cost functions study should be updated regularly to keep up with changes in the industry. b) Labor needs surveys are important, though there was some discussion of potential need for some reconfiguration of this work. c) Projections at the sub-area level are useful, but there is a need to develop projections at the local level, as well, if these studies are to be of greatest value to local community leaders and decision makers. d) Finally, it was noted that it is important to study projects through all phases of their development, and particularly not just on the front end.

3. Demographic studies were described as somewhat weak, both in the white paper and in the breakout group discussion, although ongoing studies will partially address these weaknesses. Other issues defined as important to examine in this area are in-migration, out-migration, population retention, wage pressures, and foreign immigration to meet labor demands. It was also noted that demographic changes at the local level are the major cause of public infrastructure needs.

4. Public service needs should be addressed, and are particularly important to local jurisdictions. Those identified as most important were health, police and fire protection, power, water and sewer, transportation, and education. Workshop attendees agreed that these public services should be considered in community-level analyses because of their importance at the local level. Education was defined as particularly important and is being heavily impacted by demographic changes—for example, changes in the oil and gas industry workforce have resulted in changes in curricular needs. Increasingly, labor needs demand more technical preparation, and schools are changing to address these needs. Also, as the workforce has become more ethnically diverse, new educational needs have become evident as in the need for training in English as a second language.

5. Fiscal considerations are generally covered quite well in the studies. Because these are important as drivers of local community effects, they should continue to be included in the research agenda.

6. Several studies have addressed the attitudes of local populations toward various aspects of the oil and gas industry. It has been noted, for example, that attitudes of residents in the Gulf area are changing. There is less worker loyalty, largely as a consequence of industry restructuring, layoffs, and changing labor needs. Employment in the industry is defined as less attractive as a career choice among members of the younger generation. While studies of attitudes were recognized as important, workshop participants acknowledged their limitations for the EIS process, particularly in light of OMB regulations.

7. Both the white paper and the workshop discussion concluded that sociocultural impacts are the least well-defined. At the same time, they are very important to our understanding of the local effects of the industry. Several examples bear this out: a) changes in hiring patterns are resulting in more Hispanic and Asian workers in the area. This is changing the local cultural/ethnic mix. b) A good deal of work has been done on family impacts associated with the industry. For example, several studies have addressed the effects of work patterns on family interaction and satisfaction. However, most studies have tended to focus on married male workers and their patterns of adjustment, adapting, and coping. Less attention, however, has been given to women, singles, minorities, and older workers. c) As noted in the discussion of attitudes, worker morale has changed and has been affected by changes in the industry-mergers, cost reduction initiatives, fragmentation of work teams, increased contract labor, and irregular and call-out work schedules. This affects the ability of the industry to attract and retain workers.

Several important recommendations resulted from the group discussions. These include:

1. Continuing effort must be made to make study findings a more central part of the EIS process. As currently done, the EIS process, and the GOMR socioeconomic studies activities almost occur independently of one another.

2. Efforts must also be made to make the EIS process of greater value to the communities. Community leaders and residents should be able to use the EIS findings for planning activities more directly.

3. It was agreed that there is a continuing need for industry studies to monitor change and impacts.

4. Studies are needed to address the long-term role of the oil and gas industry in the Gulf. While important historical studies have been conducted, it will be important to address longer-term industry impacts on the region.

5. Additional attention should be given to global competition. Increasingly, this is a global economy, and activities in the Gulf are significantly impacted by forces that affect the oil and gas industrial worldwide.

6. MMS should continue efforts in economic impact modeling. After years of not being state-of-the-art in this area, significant progress is now reflected in current studies. The development of more sophisticated demographic and public service components of such models should be given priority.

7. Continuing attention should be given to two difficult areas: separation of baseline and OCS-related development (it was noted in the workshop that may be an almost impossible task), and identification of specific impact areas.

8. Finally, two different approaches were reviewed and recommended in the area of impact areas: a) developing a series of comparative case studies that focus on ports; and b) directing

additional effort to monitoring local jurisdictions and more clearly identifying impacts specific to these jurisdictions.

Workshop on Deepwater Environmental Studies Strategy: Socioeconomic Breakout Group
Click on the title to view the presentation.

Workshop on Deepwater Environmental Studies Strategy:

SocioEconomic Breakout Group Overview

Co-Chairs:

Harry Luton, Minerals Management Service

Steve Murdock, Texas A&M University

Recorder:

Connie Landry, Minerals Management Service

VI. Appendices

Appendix 1. Deepwater Workshop Agenda

May 29th Morning

Welcome:	William Schroeder	8:30 – 8:40
	University of Alabama	
Introduction:	Chris Oynes, Regional Director	8:40 – 8:50
	MMS, Gulf of Mexico Region	
Industry Perspective:	David Walker	8:50 – 9:10
	BP America, Inc.	
	Paul Siegele	9:10 – 9:30
	ChevronTexaco	
DeepSpill Experiment:	Dan Allen	9:30 – 9:50
	ChevronTexaco	
Morning Break:		9:50 – 10:10
Technical Overviews:		
Physical Oceanography	Ann Jochens	10:10 – 10:30
	Texas A&M University	
Geology/Geohazards	Bill Bryant	10:30 – 10:50
	Texas A&M University	
Deep Gulf Ecology	Gil Rowe	10:50 – 11:10
	Texas A&M University	
Deep Gulf Fisheries	Randy Edwards	11:10 – 11:30
	USGS Biological Resources Division	
Socio/Economic	Steve Murdock	11:30 – 11:50
	Texas A&M University	
Closing Remarks:	William Schroeder	11:50 – 12:00

May 29th Afternoon & May 30th

Ecology Breakout Group

Co-Chairs
  Michael Rex, University of Massachusetts
  Robert Avent, Minerals Management Service
  James Ray, Shell Offshore

Recorder
Margaret Metcalf, MMS

Geology/Geohazards Breakout Group

Co-Chairs
Bill Bryant, Texas A&M University
Sarah Tsoflias, Minerals Management Service
Jim Niemann, Chevron/Texaco

Recorder
T. J. Broussard, MMS

Socioeconomic Breakout Group

Co-Chairs
Harry Luton, Minerals Management Service
Steve Murdoch, Texas A&M University

Recorder
Connie Landry, MMS

May 31st Morning

Summaries, Recommendations and General Discussion

8:45-9:00
Workshop Summaries
William Schroeder

9:00-9:20
Socioeconomics Working Group
Chair: Stan Albrecht
Harry Luton

9:20-9:40
Geology/Geohazards Working Group
Chair: Bill Bryant
Sarah Tsoflias
Jim Niemann

9:40-10:00
Ecology Working Group
Chair: Mike Rex
Robert Avent
Jim Ray

Subgroup
Chair: Jeff Childs

10:00-10:15
Break

10:15-11:30
General Discussion

11:30-11:45
Concluding Remarks
William Schroeder

Appendix 2. Breakout Group Participants

Floaters

Mary Boatman, USDOI - Minerals Management Service
Gary Goeke, USDOI - Minerals Management Service
William Schroeder, University of Alabama/Dauphin Island Sea Lab

Geology/Geophysics

Adnan A. Ahmed, USDOI - Minerals Management Service
Mary Boatman, USDOI - Minerals Management Service
T. J. Broussard, USDOI - Minerals Management Service
Bill Bryant, Texas A&M University
Robert A. George, C & C Technologies, Inc.
Ann E. Jochens, Texas A&M University
Ed Malachosky, INTEQ Drilling Fluids
James Niemann, ChevronTexaco
J. Douglas Oliver, Florida Department of Environmental Protection
Samuel Reed, Williams Energy Services
G. Ed Richardson, USDOI - Minerals Management Service
Natalia Sidorovskaia, University of Louisiana - Lafayette
Sarah L. Tsoflias, USDOI - Minerals Management Service

Physical Oceanography

Antoine Badan, CICESE
John Blaha, Naval Oceanographic Office
Donna Bourg, USDOI - Minerals Management Service
Carole Current, USDOI - Minerals Management Service
Ronald J. Lai, USDOI - Minerals Management Service
William Lang, USDOI - Minerals Management Service
Alexis Lugo-Fernandez, USDOI - Minerals Management Service
Jose' Ochoa, CICESE
Elizabeth Peuler, USDOI - Minerals Management Service
Michael Vogel, Shell Global Solutions
Michael Wild, Naval Oceanographic Office
Judy Wilson, USDOI - Minerals Management Service

Ecology

Tom Ahlfeld, USDOI - Minerals Management Service
Dan Allen, ChevronTexaco
Bob Avent, USDOI - Minerals Management Service
Robert C. Ayers, Jr., Robert Ayers & Associates, Inc.
John Blaha, Naval Oceanographic Office
Greg Boland, USOI - Minerals Management Service
Carolyn Burks, National Marine Fisheries Service
Jeff Childs, USDOI - Minerals Management Service
Susan Childs, USDOI - Minerals Management Service
Carole Current, USDOI - Minerals Management Service
George Dennis, US Geological Survey
Kevin Dischler, Sherry Laboratories
Dave Dougall, Agip Petroleum Company
Robert Dubois, US Fish & Wildlife Service
Randy Edwards, US Geological Survey
Anita George-Ares, Exxon/Mobil Biomedical Sciences Inc.
Andrew Glickman, ChevronTexaco
Alan Hart, Continental Shelf Associates, Inc.
Doug Heatwole, Ecology & Environment, Inc.
James J. Kendall, USDOI - Minerals Management Service
Jan Kenny, TEi Construction Services
Paul R. Krause, Blasland, Bouck & Lee, Inc.
Roy K. Kropp, Pacific Northwest National Laboratory
Alexis Lugo-Fernandez, USDOI - Minerals Management Service
Jamie McKee, SAIC
Margaret Metcalf, USDOI - Minerals Management Service
Paul Montagna, University of Texas at Austin
Mike Nunley, SAIC
Michael E. Parker, ExxonMobile Production Company
Jim Ray, Shell Global Solutions (US) Inc.
Larry Reitsema, Marathon Oil Company
Mike Rex, University of Massachusetts
Carol Roden, USDOI - Minerals Management Service
Bob Rogers, USDOI - Minerals Management Service
Pat Roscigno, USDOI - Minerals Management Service
Gil Rowe, Texas A&M University
Ricahrd F. Shaw, Coastal Fisheries Institute - LSU
Joseph P. Smith, ExxonMobil Upstream Research Company
Michael G. Strikmiller, Environmental Enterprises USA, Inc.
Ken Sulak, US Geological Survey
Tim Thibaut, Barry A. Vittor & Associates Inc.
Debby Tucker, Florida Department of Environmental Protection
Dick Wildermann, USDOI - Minerals Management Service
Judy Wilson, USDOI - Minerals Management Service

Socioeconomics

Stan L. Albrecht, Utah State University
Jeff Chapman, ExxonMobil
Rodney E. Cluck, USDOI - Minerals Management Service
Craig Colten, Louisiana State University
Vincent F. Cottone, ChevronTexaco
Donald W. Davis, Louisiana State University
David E. Dismukes, Louisiana State University
Warren Emerson, Conoco Inc.
Jordan B. Fiesta, University of South Florida
Roy Francis, LA1 Coalition Inc.
Gerry Gallagher, Ecology & Environment, Inc.
Stephanie Gambino, USDOI - Minerals Management Service
John E. Gibbons, John E. Gibbons Associates
Scott A. Hemmerling, Louisiana State University
Connie Landry, USDOI - Minerals Management Service
Larry Leistritz, North Dakota State University
Harry Luton, USDOI - Minerals Management Service
Brian G. Marcks, Louisiana Department of Natural Resources
Thomas R. McGuire, University of Arizona
Steve Murdock, Texas A&M University
Robin L. Petrusak, ICF Consulting
John S. Petterson, Impact Assessment, Inc.
Allan G. Pulsipher, Louisiana State University
Jeff Reidenauer, The Louis Berger Group Inc.
Claudia M. Rogers, USDOI - Minerals Management Service
Edella Schlager, University of Arizona
Joachim Singelmann, Louisiana State University
Meg Streiff, Louisiana State University
Debra Vigil, USDOI - Minerals Management Service
Barbara Wallace, TechLaw, Inc.
Dick Wildermann, USDOI - Minerals Management Service
Vicki R. Zatarain, USDOI - Minerals Management Service

Appendix 3. Workshop Attendees and Addresses

Tom Ahlfeld
Biological Oceanographer
USDOI - Minerals Management Service
381 Elden Street
Herndon VA 20170
thomas.ahlfeld@mms.gov

Adnan A. Ahmed
Geophysicist
USDOI - Minerals Management Service
1201 Elmwood Park Boulevard
New Orleans LA,70123
adnan.ahmed@mms.gov

Stan L. Albrecht
Utah State University
Provost's Office
1435 Old Main Hill
Logan UT 84322
sla@champ.usu.edu;sla@cc.lsu.edu

Dan Allen
ChevronTexaco
935 Gravier Street
New Orleans LA 70112
allj@chevron.com

Bob Avent
USDOI - Minerals Management Service
Gulf of Mexico Region
1201 Elmwood Park Boulevard
New Orleans LA 70123-2394
robert.avent@mms.gov

Robert C. Ayers Jr.
Robert Ayers & Associates, Inc.
6329 Rutgers
Houston TX 77005-3317
bobo60@aol.com

Antoine Badan
CICESE - Oceanography
Box 2732
Ensenada Mexico
abadan@cicese.mx

Eliot Barron
National Ports & Waterways Institute
1600 Canal Street, Suite 727
New Orleans LA 70112
homerule12@yahoo.com

John Blaha
Naval Oceanographic Office
N33 - 1002 Balch Boulevard
Stennis Space Center MS 39522-5001
blahaj@navo.navy.mil

Mary Boatman
USDOI - Minerals Management Service
Gulf of Mexico Region
1201 Elmwood Park Boulevard
New Orleans LA 70123-2394
mary.boatman@mms.gov

Greg Boland
USDOI - Minerals Management Service
Biological Sciences Unit
1201 Elmwood Park Boulevard
New Orleans LA 70123
gregory.boland@mms.gov

Donna Bourg
USDOI – Minerals Management Service
1201 Elmwood Park Boulevard
New Orleans LA 70123-2394
donna.bourg@mms.gov

T. J. Broussard
USDOI - Minerals Management Service
NEPA/CZM Coordination Unit
1201 Elmwood Park Boulevard
New Orleans LA 70123
tommy.broussard@mms.gov

Bill Bryant
Texas A&M University
Department of Oceanography
College Station TX 77843-3146
wbryant@ocean.tamu.edu

Carolyn Burks
National Marine Fisheries Service
3209 Frederick Street
Pascagoula MS 39567
carolyn.m.burks@noaa.gov

Jeff Chapman
ExxonMobil
800 Bell Street - Room 4111C
Houston TX 77002
jeff.s.chapman@exxonmobil.com

Susan Childs
USDOI - Minerals Management Service
GOMR/Leasing and Environment
1201 Elmwood Park Boulevard
New Orleans LA 70123
susan.childs@mms.gov

Jeff Childs
USDOI - Minerals Management Service
Environmental Sciences
1201 Elmwood Park Boulevard
New Orleans LA 70123-2394
Jeff.Childs@mms.gov

Rodney E. Cluck
USDOI - Minerals Management Service
381 Elden Street
Herndon VA 20170
Rodney.Cluck@mms.gov

Craig Colten
Louisiana State University
Geography & Anthropology
227 Howe-Russell
Baton Rouge LA 70809
ccolten@lsu.edu

Vincent F. Cottone
ChevronTexaco
Deepwater Business Unit
935 Gravier Street
New Orleans LA 70112-1625
cottovf@chevrontexaco.com

Carole Current
USDOI - Minerals Management Service
Environmental Sciences Section
1201 Elmwood Park Boulevard - MS 5433
New Orleans LA 70123
carole.current@mms.gov

Donald W. Davis
Louisiana State University
LA Applied & Educ. Oil Spill R&D Program
258A/B Military Science Building
Baton Rouge LA 70803
osradp@attglobal.net

George Dennis
USGS
7920 NW 71st Street
Gainesville FL 32653
george_dennis@usgs.gov

Kevin Dischler
Sherry Laboratories-Louisiana
Bioassay Division
2417 West Pinhook Road
Lafayette LA 70508
kevind@sherrylabs.com

David E. Dismukes
Louisiana State University
Center for Energy Studies
Baton Rouge LA 70803-0301
dismukes@lsu.edu

Dave Dougall
Agip Petroleum Co
1201 Louisiana - Suite 3500
Houston TX 77002
david.dougall@agippetroleum.agip.it

Robert Dubois
U.S. Fish & Wildlife Service
646 Cajundome Boulevard - Suite 400
Lafayette LA 70506
robert_dubois@fws.gov

Randy Edwards
USGS
College of Marine Science
600 Fourth Street South
St Petersburg FL 33701-4846
redwards@usgs.gov

Joan Elterman
Dominion Exploration & Production Inc.
1450 Poydras Street
New Orleans LA 70112
joan_a_elterman@dom.com

Warren Emerson
Conoco Inc.
Magnolia
Houston TX 77252
Warren.Emerson@conoco.com

Caryl Fagot
MMS - ORD
1201 Elmwood Park Boulevard
New Orleans LA 70123
caryl.fagot@mms.gov

Jordan B. Fiesta
University of South Florida
2664 Champion Ridge Drive
Lakeland FL 33813
Fiesta@helios.acomp.usf.edu

Roy Francis
LA1 Coalition Inc.
NSU
PO Box 2048
Thibodaux LA 70310
slec-rpf@nicholls.edu

Sandi Fury
ChevronTexaco
935 Gravier Street
New Orleans LA 70112
sfur@chevrontexaco.com

Gerry Gallagher
Ecology & Environment Inc.
1950 Commonwealth Lane
Tallahassee FL 32303
gagallagher@ene.com

Stephanie Gambino
USDOI - Minerals Management Service
LE-EA-NEPA/CZM
1201 Elmwood Park Boulevard - MS 5412
New Orleans LA 70123
stephanie.gambino@mms.gov

Robert A. "Tony" George
C & C Technologies Inc.
Geosciences
730 E. Kaliste Saloom Road
Lafayette LA 70508
tony.george@cctechnol.com

Anita George-Ares
ExxonMobil Biomedical Sciences Inc.
Occupational & Public Health
1545 Route 22 East
PO Box 971
Annandale NJ 08801-0971
anita.george-ares@exxonmobil.com

David Gettleson
Continental Shelf Associates Inc.
759 Parkway Street
Jupiter FL 33469
dgettleson@conshelf.com

John E. Gibbons
John E. Gibbons Associates
1115 Ranch Point Way
Antioch CA 945310
JEGCA@aol.com

Andrew Glickman
ChevronTexaco
Energy Research and Technology Co.
100 Chevron Way
Richmond CA 94802
aglickman@chevrontexaco.com

Gary Goeke
Ecology and Environment, Inc.
220 West Garden Street, Suite 404
Pensacola, FL 32501
ggoeke@ene.com

Alan Hart
Continental Shelf Associates Inc.
759 Parkway Street
Jupiter FL 33477
ahart@conshelf.com

Doug Heatwole
Ecology & Environment Inc.
220 W. Garden Street - Suite 404
Pensacola FL 32501
dheatwole@ene.com

Scott A. Hemmerling
Louisiana State University
Geography & Anthropology
539 Park Boulevard #3
Baton Rouge LA 70806
shemme1@lsu.edu

Ann E. Jochens
Texas A&M University
Department of Oceanography
3146 TAMU
College Station TX 77843-3146
ajochens@tamu.edu

James J. Kendall
USDOI - Minerals Management Service
381 Elden Street - Mail Stop 4041
Herndon VA 20710-4817
james.kendall@mms.gov

Jan Kenny
TEi Construction Services - DEMEX Division
720 Grefer Avenue
Harvey LA 70058
demex-nola@msn.com

John Kenny
TEi Construction Services - DEMEX Division
702 Grefer Avenue
Harvey LA 70058
demex-nola@msn.com

Robert King
Kvaerner Oilfield Products
1255 N Post Oak Road
Houston TX 77055
robert.king@kvaerner.com

Paul R. Krause
Blasland Bouck & Lee Inc.
Life Sciences
301 E Ocean Boulevard - Suite 1530
Long Beach CA 90802
pkrause@bbl-inc.com

Roy K. Kropp
Pacific Northwest National Laboratory
Marine Sciences Laboratory
1529 West Sequim Bay Road
Sequim WA 98382
roy.kropp@pnl.gov

Ronald J. Lai
USDOI - Minerals Management Service
381 Elden Street
Herndon VA 20170
ronald.lai@mms.gov

Connie Landry
USDOI - Minerals Management Service
Gulf of Mexico OCS Region
1201 Elmwood Park Boulevard
New Orleans LA 70123-2394
connie.landry@mms.gov

William Lang
USDOI - Minerals Management Service
1201 Elmwood Park Boulevard
New Orleans LA 70123
bill.lang@mms.gov

Larry Leistritz
North Dakota State University
Fargo ND 58105
lleistri@ndsuext.nodak.edu

Alexis Lugo-Fernandez
USDOI - Minerals Management Service
1201 Elmwood Park Boulevard
New Orleans LA 70123-2394
alexis.lugo.fernandez@mms.gov

Harry Luton
USDOI - Minerals Management Service
Gulf of Mexico Region
1201 Elmwood Park Boulevard
New Orleans LA 70123-2394
harry.luton@mms.gov

Ed Malachosky
INTEQ Drilling Fluids
2001 Rankin Road
Houston TX 77073
ed.malachosky@inteq.com

Brian G. Marcks
Louisiana Department of Natural Resources
Coastal Management Division
617 N 3rd Street
Baton Rouge LA 70802
BrianM@dnr.state.la.us

Jonathan Martin
Dauphin Island Sea Lab
101 Bienville Boulevard
Dauphin Island AL 36528
jmartin@disl.org

Thomas R. McGuire
University of Arizona
Bureau of Applied Research in Anthropology
Tucson AZ 85621
mcguire@u.arizona.edu

Jamie McKee
SAIC
1140 Eglin Parkway
Shalimar FL 32579
mckeew@saic.com

Margaret Metcalf
USDOI - Minerals Management Service
Leasing and Environment
1201 Elmwood Park Boulevard
New Orleans LA 70123-2394
Margaret.Metcalf@mms.gov

Paul Montagna
University of Texas at Austin
Marine Science Institute
750 Channel View Drive
Port Aransas TX 78373
paul@utmsi.utexas.edu

Steve Murdock
Texas A&M University
Department of Rural Sociology
Building E - Suite 100
College Station TX 77843-2125
smurdock@tamu.edu

James Niemann
ChevronTexaco
Deepwater Business Unit
935 Gravier Street
New Orleans LA 70112
jamesniemann@chevrontexaco.com

Mike Nunley
SAIC
1140 Eglin Parkway
Shalimar FL 32579
nunleyj@saic.com

Jose' Ochoa
CICESE
Oceanography Research
Mm 107 Carr- Tijuana-Ensenada
Baja California Mexico 22800
jochoa@cicese.mx

J. Douglas Oliver
Department of Environmental Protection
Office of Intergovernmental Programs
3900 Commonwealth Boulevard - MS-47
Tallahassee FL 32399-3000
Doug.Oliver@dep.state.fl.us

Chris Oynes
USDOI - Minerals Management Service
Gulf of Mexico Region
1201 Elmwood Park Boulevard
New Orleans LA 70123-2394
chris.oynes@mms.gov

Michael E. Parker
ExxonMobil Production Company
800 Bell Street - Room 4289
Houston TX 77002
michael.e.parker@exxonmobil.com

Robin L. Petrusak
ICF Consulting
9300 Lee Highway
Fairfax VA 22031
robinpetrusak@icfconsulting.com

John S. Petterson
Impact Assessment Inc.
2166 Avenida de la Playa - Suite F
La Jolla CA 92037
iai@san.rr.com

Elizabeth Peuler
Physical Sciences Unit
USDOI - Minerals Management Service
1201 Elmwood Park Boulevard
New Orleans LA 70123
elizabeth.peuler@mms.gov

Allan G. Pulsipher
Louisiana State University
Center for Energy Studies
One East Fraternity Circle
Baton Rouge LA 70803
agpul@lsu.edu

Jim Ray
Shell Global Solutions (US) Inc.
Westhollow Technology Center
3333 Highway 6 South
Houston TX 77082
james.ray@shell.com

Samuel Reed
Williams Energy Services
2800 Post Oak Boulevard
Houston TX 77056
samuel.reed@williams.com

Jeff Reidenauer
The Louis Berger Group Inc.
100 Halstead Street
East Orange NJ 07018
jreiden@louisberger.com

Larry Reitsema
New Ventures and Technical Support
Marathon Oil Corporation
5555 San Felipe
Houston TX 77056
lareitsema@marathonoil.com

Mike Rex
University of Massachusetts
Department of Biology
100 Morrissey Boulevard
Boston MA 02125-3390
michael.rex@umb.edu

G. Ed Richardson
USDOI - Minerals Management Service
Leasing and Environment
1201 Elmwood Park Boulevard
New Orleans LA 70123-2394
ed.richardson@mms.gov

Carol Roden
USDOI - Minerals Management Service
Leasing and Environment
1201 Elmwood Park Boulevard
New Orleans LA 70123
carol.roden@mms.gov

Claudia M. Rogers
USDOI - Minerals Management Service
817 Phlox Avenue
Metairie LA 70001
claudia.rogers@mms.gov

Bob Rogers
Biological Sciences Unit
USDOI - Minerals Management Service
1201 Elmwood Park Boulevard
New Orleans LA 70123
Robert.Rogers@mms.gov

Pat Roscigno
USDOI - Minerals Management Service
1201 Elmwood Park Boulevard
New Orleans LA 70123
pasquale.roscigno@mms.gov

Gil Rowe
Texas A&M University
Department of Oceanography
College Station TX 77843-3146
growe@ocean.tamu.edu

Edella Schlager
University of Arizona
Tucson AZ 85721
eschlager@bpa.arizona.edu

Randy Schlude
Dauphin Island Sea Lab
101 Bienville Boulevard
Dauphin Island AL 36528
rschlude@disl.org

William Schroeder
The University of Alabama
Dauphin Island Sea Lab
101 Bienville Boulevard
Dauphin Island AL 36528
wschroeder@disl.org

Cheryl Sephus
University of New Orleans
Department of Physics
UNO PO BOX 1014
New Orleans LA 70148
midsea20@hotmail.com

Richard F. Shaw
Coastal Fisheries Institute - LSU
Oceanography & Coastal Science
Wetlands Resources Building
Baton Rouge LA 70806
rshaw@lsu.edu

Natalia Sidorovskaia
University of Louisiana - Lafayette
Department of Physics
UL BOX 44210
Lafayette LA 70503
nsidorovskaia@louisiana.edu

Paul Siegele
ChevronTexaco
935 Gravier Street
New Orleans LA 70112

Joachim Singelmann
Louisiana State University
Department of Sociology
Baton Rouge LA 70803
joachim@lsu.edu

Joseph P. Smith
ExxonMobil Upstream Research Co
Offshore Division
3616 Richmond
Houston TX 77046
joe.p.smith@exxonmobil.com

Meg Streiff
Louisiana State University
Center for Energy Studies
2543 Hundred Oaks
Baton Rouge LA 70808
mstrei1@lsu.edu

Michael G. Strikmiller
Environmental Enterprises USA Inc.
58485 Pearl Acres Road - Suite D
Slidell LA 70461
mstrikmiller@eeusa.com

Ken Sulak
USGS
Biological Resources
7920 NW 71st Street
Gainesville FL 32653
ken_sulak@usgs.gov

Tim Thibaut
Barry A. Vittor & Associates Inc.
8060 Cottage Hill Road
Mobile AL 36695
tthibaut@bvaenviro.com

Sarah L. Tsoflias
USDOI - Minerals Management Service
1201 Elmwood Park Boulevard - MS 5433
New Orleans LA 70123-2394
Sarah.Tsoflias@mms.gov

Debby Tucker
Department of Environmental Protection
Office of Intergovernmental Programs
3900 Commonwealth Boulevard - MS-47
Tallahassee FL 32399-3000
Debby.Tucker@dep.state.fl.us

Debra Vigil
USDOI - Minerals Management Service
1201 Elmwood Park Boulevard
New Orleans LA 70123
debra.vigil@mms.gov

Michael Vogel
Shell Global Solutions
3737 Bellaire Boulevard
Houston TX 77025
michael.vogel@shell.com

David Walker
British Petroleum
501 Westlake Park Blvd
Room 20.186
Houston TX 77079
walkerdb@bp.com

Barbara Wallace
TechLaw Inc.
4340 East West Highway - Suite 1120
Bethesda MD 20814
bwallace@techlawinc.com

Michael Wild
Naval Oceanographic Office
Stennis Space Center MS 39525
wildm@navo.navy.mil

Dick Wildermann
USDOI - Minerals Management Service
Branch of Environmental Assessment
381 Elden Street
Herndon VA 20171
richard.wildermann@mms.gov

Judy Wilson
USDOI - Minerals Management Service
381 Elden Street
Herndon VA 20170
judy_wilson@mms.gov

Carolyn Wood
Dauphin Island Sea Lab
101 Bienville Boulevard
Dauphin Island AL 36528
cwood@disl.org

Vicki R. Zatarain
USDOI – Minerals Management Service
1201 Elmwood Park Boulevard
New Orleans LA 70123
vicki.zatarain@mms.gov

The Department of the Interior Mission

As the Nation's principal conservation agency, the Department of the Interior has responsibility for most of our nationally owned public lands and natural resources. This includes fostering sound use of our land and water resources; protecting our fish, wildlife, and biological diversity; preserving the environmental and cultural values of our national parks and historical places; and providing for the enjoyment of life through outdoor recreation. The Department assesses our energy and mineral resources and works to ensure that their development is in the best interests of all our people by encouraging stewardship and citizen participation in their care. The Department also has a major responsibility for American Indian reservation communities and for people who live in island territories under U.S. administration.

The Minerals Management Service Mission

As a bureau of the Department of the Interior, the Minerals Management Service's (MMS) primary responsibilities are to manage the mineral resources located on the Nation's Outer Continental Shelf (OCS), collect revenue from the Federal OCS and onshore Federal and Indian lands, and distribute those revenues.

Moreover, in working to meet its responsibilities, the **Offshore Minerals Management Program** administers the OCS competitive leasing program and oversees the safe and environmentally sound exploration and production of our Nation's offshore natural gas, oil and other mineral resources. The MMS **Minerals Revenue Management** meets its responsibilities by ensuring the efficient, timely and accurate collection and disbursement of revenue from mineral leasing and production due to Indian tribes and allottees, States and the U.S. Treasury.

The MMS strives to fulfill its responsibilities through the general guiding principles of: (1) being responsive to the public's concerns and interests by maintaining a dialogue with all potentially affected parties and (2) carrying out its programs with an emphasis on working to enhance the quality of life for all Americans by lending MMS assistance and expertise to economic development and environmental protection.

www.ingramcontent.com/pod-product-compliance
Lightning Source LLC
Chambersburg PA
CBHW082302200526

45168CB00017B/2492